The Projectionists

Stephen Barber

The Projectionists

Eadweard Muybridge and the Future Projections of the Moving Image

DIAPHANES

Supported by a Publications Grant from the
Paul Mellon Centre for Studies in British Art

1st edition

ISBN 978-3-0358-0289-4

Layout and prepress: 2edit, Zurich
Printed in Germany

www.diaphanes.com

Contents

Acknowledgements

I am very grateful to the Paul Mellon Centre for Studies in British Art, the Leverhulme Trust and the Great Britain Sasakawa Foundation for generously supporting this book, to Becky Beasley, Patrick ffrench, Stephen Herbert, Xavier Ribas and Michael Witt, and especially to Seoyoung Kim, Curator of the Eadweard Muybridge Collection at the Kingston Museum. I would also like to thank the staff – particularly Stefan Donath, Holger Hartung and Torsten Jost – at the Interweaving Performance Cultures international research centre at the Free University Berlin, where this book was partly written, and also my colleagues at the Kingston School of Art, Kingston University, London.

The Projectionists

The art of moving-image projection forms a volatile entity whose most prescient experiments – from their 1890s origins to the contemporary moment, from glass discs to celluloid to the digital image – vitally envision the future of the human body and of the spaces of cities, but are also striated by contamination, disintegration and obsolescence. As well as a seminal innovator of panoramic photography, landscape photography and motion photography, Eadweard Muybridge, in the final years of his work, was also the moving image's first projectionist, travelling incessantly from city to city to project his sequences to awestruck beyond-capacity audiences and finally constructing the first purpose-designed urban auditorium conceived solely for the projection of moving images, before his work abruptly fell into catastrophe.

Projection is also pivotally an obsession, and the corporeal figure of the moving-image projectionist is marked by obsessive addictions determined by film's engulfing or haywire time. I learned the art of moving-image projection at the age of seventeen, from an elderly master projectionist, Mr Schultz, who had begun his work in the great "cinema-palaces" of 1920s Vienna and Berlin, had projected across the cities of the Soviet Union as a travelling projectionist, from Tallinn to Magadan and from Lvov to Tbilisi, in the following decade, had kept on projecting in desperate obstinacy in Baltic coastal cities' cinemas left standing isolated in wastelands by firestorms that had otherwise consumed entire districts of those cities, before finally escaping from the port of Stettin and pursuing his occupation as a stateless refugee in the North of England over the postwar decades; he was now the senior projectionist at the Plaza cinema in Leeds. By the time he taught me projection's imperatives, at his workshop in an industrial building alongside the river Aire in 1979, he was seventy-five years old but as obsessed as ever, unwilling to retire, the deep pockets of his pre-war leather overcoat crammed with rivets, screws, claws, springs, junctures and scalpels that could suture any celluloid rip or resuscitate any projector's malfunction. He gestured with his worn, scarred fingers and massaged the fissured skin around his exhausted eyes as he spoke with passion for hours about the nuances of the beater mechanism, the Maltese cross, the aperture mask, the hidden iris. He taught me the intensive concentration needed for performing split-second reel-changeovers, the choreography of oscillation required in switching adeptly between two projectors, the necessary near-psychic anticipation of celluloid's tearings, and the suspension of blind panic demanded when a projector caught fire or the screen suddenly went blank or the sound lapsed and the spectators began to scream and roar. He had fought innumerable battles over fifty or more years with the audiences of cinemas who had besieged his projection-box's door in fury at seemingly irreparable projection failures, or even in protest at the programme of films he was projecting. He had projected everything, from Murnau to pornography films, from exploitation films to Lang. At the end of the evening, he would drink a few glasses of a strong Ukrainian spirit and begin to reminisce about his travels of the 1920s and 1930s, until his son, who had heard it all before, rolled his eyes and we would slip away. His son was my age, and had been assigned by his father to the exacting occupation of a relief projectionist, travelling most evenings in a battered van to local cinemas across the entire terrain of Yorkshire, from Thirsk to Penistone, from Hardraw Scar to Spurn Head Spit, to substitute for any cinema's projectionist who had fallen ill or had vanished. A cinema's manager would telephone the old projectionist, and he would note down the location and the number of hours' work at stake, negotiating the cost after assessing the extent of fear in the manager's voice. Especially in the summertime, that relief occupation expanded, and on one occasion when I visited the old projectionist's workshop with his son, he asked me if I wanted some work. He would take a third of my fees, in return for the bookings and for showing me how the job worked

("I'll initiate you", he told me). I had time on my hands before starting art college, and so I received my training from him and spent the summer travelling from decrepit cinema to cinema, projecting pornography, *Apocalypse Now*, *Zulu Dawn*, *Nosferatu*, whatever was required. If the booking was an informal one from a club or society for enthusiasts, the film they wanted to see that summer was usually *Stalker*, projected from directly behind the audience in a schoolroom or canteen. The cinemas often appeared last-ditch, last-gasp places, barely surviving, the Bell & Howell and Kinoton projectors last maintained decades ago. When I arrived for my shifts, the managers often eyed me with unease, as the counterfeit stand-in to the master projectionist's son, wondering if I was up to the job, but I had been taught well, and even under pressure – when a reel split-apart or a projector ground to a halt – I somehow kept going, honing the reel changeovers to a hair's-breadth. Often, when I glanced down through the projection-box window into the auditorium at the film's end, nobody remained there, and even the manager and cashier had already left. I kept travelling all summer, needing to be at the designated cinema at least an hour before the film started in order to resolve any technical issues before the projection, over the highways of Yorkshire, the M1, the A1, the M62, through the mountains and across the bleak moors, from city to city, alongside the steam-exhaling power-stations and over the rivers, from cinema to cinema. Apart from one time when I accompanied the old projectionist's son to what was anticipated to be a challenging job with potentially incendiary projectors, I always worked alone, in solitude, barely exchanging a word with anyone throughout that summer – since the old projectionist dealt with the fees, there was never any need for transactions with the cinemas' managers. Soon, I was habituated entirely to the autonomous and mysterious spatial environment of projection, with its always-imminent disasters and crises.

I left the employment of the old projectionist at the summer's end to start art college. Mr Schultz evidently believed I had what it took to become a professional projectionist, but he was ready to dismiss me anyway, now that the harsh Yorkshire winter was about to come down and there would be barely enough relief work for his son alone, sliding over the roads' ice-sheets in his van to reach marooned cinemas and project films to desperate audiences as though he were a mountain doctor. Mr Schultz gave me the last of my filmic earnings after taking his lavish cut, then shook my hand with his own calloused, multiply nicked and scarred hand. Even before I had left his workshop, he was back absorbed at his workbench, eyes focused, manipulating a projector's defective shutter.

I took the art college's course in experimental film that was optional for first year students, then specialised in it in the second and third years. Half of the time was devoted to watching or making experimental films, half of the time to film theory; Laura Mulvey was one of the theory tutors, and experimental filmmakers often arrived from Europe and the USA to give talks on their work. The college had a small screening-room with around thirty seats, with two 16mm projectors and an 8mm projector in the projection-box. The college employed no projectionist, and for all film projections, the students had to serve as their own projectionists, taking it in turns, unless any of them was so possessed by the desire to project, to the point of obsession, that they seized the turn of anyone who hesitated. A technician gave some basic induction talks, then the projections began. For months on end, we watched Stan Brakhage, Maya Deren, Kurt Kren, Kenneth Anger and James Benning films. Often, the more extended films such as those of Gregory Markopoulos and Warhol lasted late into the night, the projectionist and avid audience wired on amphetamines. But many films lasted only a few seconds, and often required immense skill to project, especially when the celluloid arrived coated in delicate insects' wings, or with its surface ripped by razors, or intentionally subjected to deterioration. Then, the projectionist's hands manoeuvred the film towards the projector gate as a precious artefact that demanded intense care, or would oth-

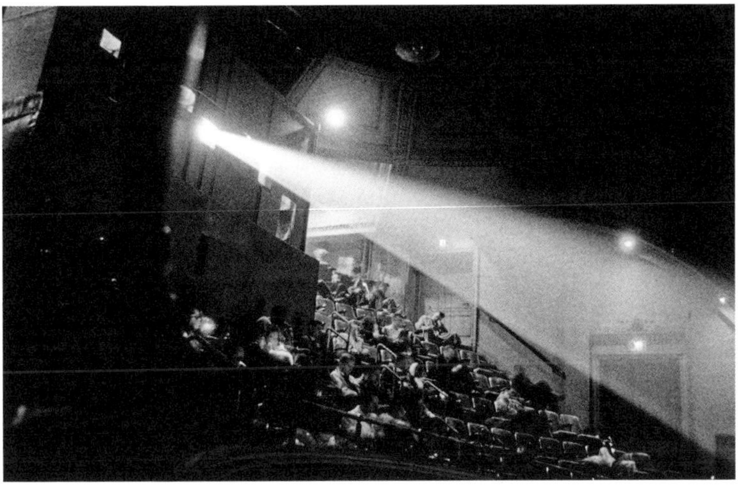

Ill. 1. *42nd Street movie theater audience, N.Y.C.*, 1958.

erwise be ready to fall apart into celluloid's death. The tactile sense of those films' durations and immediacies, of their fragile skins and depths, grew most explicit whenever they were revivified through projection. We also occasionally projected feature-films all through the night in the college's canteen for any of the art students who wanted to attend – films by Lynch, Tarkovsky, Russell, Waters, Cronenberg – with audiences of 200 or so, and for the first time, especially when the hired 35mm projector malfunctioned or else through narcotic intensity, I experienced the riotous wild uproar and violence of all-night screenings, the screams and hallucinatory deliria, the attacks and projectiles launched at the screen, then the audience transformed into comatose figures at dawn.

I have encountered the projectionists, in many countries, cities and cinemas. Over the course of arts research into cinema architectures, I traversed the moment in the second half of the 1990s when the profession of the celluloid projectionist abruptly fell into danger, obsolescence and disappearance, after a century of existence. Even at what appeared to be the most abandoned cinemas – abandoned already for decades – there was always a projectionist hanging around somewhere. At some point in my exploration of a cinema's architecture, I would reach the projection-box, and knock on the door, knowing never to enter the projectionist's domain unannounced. Someone inside would curse in livid alarm as though a painstaking momentum of concentration had been shattered, but on most occasions, the door would finally open and I would be scrutinised by eyes more habituated to celluloid strips and projection lenses than to other eyes. If I could identify and remark on an unusual projection set-up, or somehow intimate that I, too, in my own way, was a projectionist, the ice was broken, otherwise that encounter would be glacial.

Projectionists, over moving-image projection's history, were almost invariably addicts, either of film or of a narcotic intake demanded for perpetual maximal concentration, often to be sustained overnight or even, in situations of precarious or exploitative labour, over several consecutive days. Sleeplessness itself is a staring-eyed addiction in which future days of film are ready to be consumed raw, and new, near-skeletal and organ-less corporealities emerge. The risk of an eventual far-gone disintegration of an addicted projectionist would immediately manifest itself in slipshod film-focus, off-kilter changeovers between projectors, and awry sound, all invariably generating the rage of the audience and then of the cinema manager. As a result, the addiction of the projectionist forms a finetuned, static one. The projectionist is, above all, a professional perfectionist, coldly abhorring amateurs, part-timers, itinerants, or gushing enthusiasts. Whenever projection is undertaken in experimental forms, from Muybridge's 1891 European projection-tour to contemporary digital immersive projections, the demands on the projectionist are infinitely multiplied.

I spent a year in Los Angeles' Downtown Broadway avenue, researching its twelve spectacular cinemas that had been constructed between 1908 and 1928, including the Tower Theatre cinema whose auditorium had been filmed by David Lynch in *Mulholland Dr.*, and the Million Dollar Theatre cinema whose illuminated facade appears in *Blade Runner*. That avenue of cinemas had been at the heart of Los Angeles' filmic core for several mid-century decades, presenting grand premieres of Chaplin's films and lavish programmes for capacity audiences that arrived there from throughout the city by red and yellow streetcars. By 2008, none of those cinemas any longer presented regular programmes of films, but all still had a projectionist. In one, now-ruined cinema with a torn screen, a projectionist had to be retained and ready in the part-collapsed projection-box in case a potential multi-million dollar buyer for the building casually requested a projection, in another, its auditorium now used as an evangelist church, the projectionist beamed out slides of religious invocations, in another, now used for the storage of valuable sound-equipment, the projectionist's eyes had become attentive organs of surveillance, and in another, now used as a wild nightclub, the projectionist projected strobed sequences mixing pornography and animation directly onto the clientele's bodies to accentuate their frenzy. In one or two cinemas, such as the Los Angeles Theatre, the projectionists retained their sense of professional dignity, projecting 1920s films for festivals or new films at test-screenings. Once the Los Angeles Theatre's projectionist had got to know me, he would invite me into his projection-box to demonstrate the acutely adept speed of his work, honed for a half-century, his eyes perpetually in movement from the reels' revolutions through the gate, across the audience's near-vertical rake of seating, towards the immense screen itself, and back; an intricate system of vertical mirrors allowed his work simultaneously to be projected to the clientele of a lounge in the cinema's basement. He worked always with addicted passion, but I could see his face furrowed with anxiety. Finally, he told me he had a competitor, another master projectionist with a half-century's experience, who now loitered incessantly on the starburst mosaic pavement directly outside the cinema. When the cinema's workload had diminished a few years ago to the extent that only one of the two projectionists was still required, the manager had simply flipped a coin and dismissed one of them, so that set-aside projectionist became a lost filmic corporeality of addicted Los Angeles, cast astray. The surviving projectionist told me that he feared above all a reversal of that tossed coin, as though in a hallucination or an aberration of memory, so that the cast-aside projectionist would suddenly hammer on the projection-box's door, in mid-projection, back from projection's death-zone, now demanding his rightful place back.

In many countries across Europe, I also often met the projectionists of insane asylums, usually drawn from the ranks of the inmates, and often women (while I less often met women projectionists working in cinemas), projecting with an addicted focus only enhanced by multiple variants of psychosis in those asylums' often-majestic ballrooms, the patients in their many hundreds brought in and seated on lines of plastic chairs. In such environments, the collections of celluloid films available to be projected had been accumulated haphazardly over many decades, but the audiences' engrained limb-spasms and cacophonic vocal dual-deliria would be stilled utterly as soon as the auditorium's lights went down and the projector's illuminated beam traversed the space above their heads, to project Powell's *Peeping Tom* or Dreyer's *The Passion of Joan of Arc* (a film which only survived in its complete form through its projection's prevalence in 1930s insane asylums). In countries with warm summer nights, outdoor cinemas in asylum grounds appeared widely in the 1950s at the beginnings of an era of anti-psychiatric experimentation in asylum film-projections, during which the insane watched films of surrealism; often, those open-air cinemas installed among the asylum parklands' trees were constructed from one great pre-cast assemblage of steel, formed of seat-rows with the projection-box welded behind. When darkness fell and the patients had been assembled and seated by the orderlies, their glossolalia would suddenly quieten as the figure of the asylum's projectionist emerged mysteriously from between the cypress trees, and the projection could begin: exposed cinemas of madness, with projections for and by the mad.

The eminent Tokyo-based film-historian Donald Richie knew all of the city's projectionists and was always eager to introduce me to them on our 1990s walks through that city, from the gravely focused master projectionists of Shinjuku district's immense auditoria such as the Koma cinema to the suave young art-projectionists of Tokyo's avant-garde cinemas such as Uplink Factory, and from the ultimately professional projectionists of the National Film Theatre to the harassed projectionists of Tokyo's last-surviving pornography cinemas in Ueno district, who complained of suffering obstructions in their attempts to check their projections' focus on the screen since the rear of the auditorium was often the location of compulsive sexual activity that could block the projection-box's window. Richie, who was an experienced projectionist himself from his 1960s work as an experimental filmmaker (often needing to project his own films at their screenings), commented wryly that many lubricants were always vital in a projection-box, to avert jammings and stallings, and at least here, if nothing else was available, semen would do.

Bangkok had possessed among the greatest number of cinemas of any megalopolis at its filmic zenith, but when I visited the city in the 2000s to explore its surviving cinemas' architecture, its detrital projectionists had become downcast figures. Aged and exhausted but determined to maintain their filmic addictions, they spoke with desolate passion about the moment during the Vietnam war when Bangkok became the optimum destination for US military leave, and thriving cinemas suddenly proliferated on Sukhumvit avenue with names drawn from European and US cities: London cinema, Washington cinema. The London cinema's projection-box was pitched high above the vast auditorium and could be reached only via a precarious metal ladder, as with that of the Patria cinema in the Moldovan city of Chisinau; the London cinema still operated once a week, its facade's 1960s neon signage momentarily illuminated at the dead of night, but I could see that its decrepit, depopulated auditorium would imminently be eviscerated, that space transformed as with those throughout Bangkok into a car-parking shed, leaving that cramped projection-box and its occupant marooned in mid-air.

The obsolescence of the projectionists accelerated across the years from the mid-1990s, since to project a digital film often required only a pre-programmed activation and a cursory check of the screen's pixilation ratios. But many cinemas worldwide obstinately persisted with celluloid projection, at least in part through the obsessional tenacity of the projectionist's figure. (Some cinemas, such as those of art museums and national film centres, would sustain that figure of the celluloid projectionist indefinitely.) Any cinema's final relinquishing of celluloid, and its conversion to digital projection, became an increasingly rare and momentous event in the 2010s, as with that of Berlin's Filmtheater am Friedrichshain cinema, where Joseph Goebbels, always film-enthralled, had given speeches in the 1930s, and had stood beside the screen vocally to introduce and comment on his favoured films. I had met the cinema's elderly, adept projectionist on several occasions, and once he had projected the final celluloid reel of Wong Kar-Wai's film *The Grandmasters* for the last time, I saw him emerge into the foyer weeping at celluloid's loss, then vanish into the darkened street outside. He was himself a projectionist grandmaster.

The figure of the projectionist – restlessly itinerant or else inhabiting a carefully controlled spatial domain, and often focused on projected film's multiple deaths and contaminations as well as upon its upheavals of corporeality and urban space – appears in many feature films, but only two or three films seize the pivotal obsessions at stake in the projectionist's work. The projectionist's figure in film is invariably the envisaging of an isolated one, in which an encounter with another figure allows those obsessions to become openly malleable for a moment, until the projectionist's isolation closes down once again.

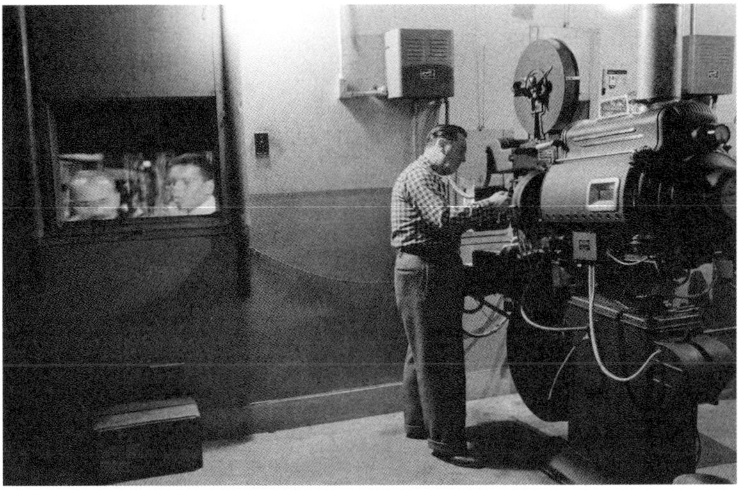

Ill. 2: *Man working with a projector*, 1958.

In Powell's *Peeping Tom* (1960), the projectionist Mark has constructed his own expansive projection-box in the attic of his large house, that fiercely protected filmic zone now inhabiting the workshop in which his child-psychologist father once professionally observed fear, deploying his son's eyes as apertures into fear's study. Mark's own professional status is that of a cinematographic assistant at a film studio, but he is also a professional killer, piercing the throats of his victims with a spiked tripod as he approaches them to film their wide-eyed terror, thereby honing his father's work. Mark's acts of killing and his lethal projections appear to form an unendable sequence, seemingly accomplishable only if he were also able to dissect and project the image of his own approaching face (and film-camera lens) from the retinas of his victims, as the German experimental anatomist Wilhelm Kühne attempted to do in 1878 – around the same moment as Muybridge's first experiments in projection – through his science of optography, by recording the last-moment content of a rabbit's eye before its death, and with the unrealized aim to dissect and project the image imprinted upon a human eye's retina at the moment before its owner's violent killing. Finally, and above all, Mark is a professional projectionist whose projection-box and screening-room hold no intervening wall, so that he projects his sequences to himself, in intimacy with his screen, until he convulses in orgasm. Two figures successively pierce the isolation of his projection-room, his young female tenant Helen and then her blind mother, and in each instance he is compelled, as a master projectionist, to project his films to them, even though Helen may betray his filmic killings to the police, and her mother, in her blind vision, is able to intuit the passage of Mark's films between projector and screen, and thereby also endanger the sustaining of his killings and their projections. Helen and Mark take a walk together through the city, and it appears that Mark's obsessional attachment to his projections may break, but Helen finally incenses him by usurping his projectionist's position, projecting one of his films of death to herself during his absence, until he catches her in that act and calamity ensues.

When Mark knows that the police are arriving to arrest him, he performs his own death as an act of Muybridge-inspired motion-photography which he has meticulously prepared in advance, anticipating his own death. Mark uses the system of feet-triggered multiple cameras – arranged sequentially along one side of the course of his running body, their shutters activated a fraction of a second apart – which Muybridge had deployed in order to capture with extreme accuracy the running movement of horses, most notably in 1881 at his patron Leland Stanford's farm in Palo Alto, California (Muybridge then adapted those gridded sequences into spinning glass-disc media for his subsequent moving-image projection tours). Once Mark has traversed that headlong space of death's recording, he finally pierces his own throat with his spiked tripod, filming his own eyes as he dies. Helen weeps over his body, but she has already demonstrated her own aptitude as a projectionist, and beyond the film's end, she will assemble that motion-photography sequence, together with the fragment of film revealing Mark's eyes in death, and project them.

In Wenders' *Kings of the Road* (1976), a repairer of cinemas' ailing, last-ditch projectors drives and also inhabits a lorry crammed with projector parts as he travels between small towns' cinemas along West Germany's border with East Germany, north to south. Although Bruno is itinerant and his work is framed by intense solitude, he has a prepared schedule for that itinerancy, which he will tear apart at the film's end. He picks up a child-psychologist who has driven his car into the Elbe river in desolation, and they travel together for a time, but each remains locked in isolation. Cinema projection is vanishing and the cinemas will not survive; Bruno's work is to maintain the cinemas' projectors just long enough so that they can transit over into film's death. Although Bruno is a sustainer or resuscitator of projectors and not himself a professional projectionist, he knows projection's demands intimately and is exasperated whenever he encounters the small towns' inane projectionists, who know almost nothing of their profession, having never heard of the projector's Maltese cross.

On one occasion, Bruno is forced himself to incorporate the figure of the projectionist. He attends a pornographic projection of such ineptitude that he is compelled to infiltrate his way from the auditorium into the cinema's projection-box, and discovers the relief projectionist now masturbating, having obstructed part of his projection beam so that it deflects backwards across the space of the projection box, onto an improvised miniaturised second screen on the back wall; as a result, the masturbating projectionist is obliviously facing away from the cinema's screen, and does not care anyway about the auditorium's malfunctioning projection. Bruno adeptly manoeuvres the obstructing device so that the secondary projection beam is now propelled directly into the face of the disgruntled projectionist, pornography's frenzied corporeal amalgams punitively burned into his facial skin and his eyes, rendering him enraged, so that he abandons entirely his half-hearted profession and storms out of the cinema in mid-projection. Bruno must then momentarily assume that professional role, rectifying the askew projection and gathering up the mass of celluloid which the vanished projectionist had let uncoil onto the projection-box floor.

Kings of the Road opens and closes with sequences that vocally reflect upon the status of projection and the survival of the figure of the projectionist. In the first sequence, an elderly male cinema owner recounts his personal history since the 1920s while Bruno is engaged in repairing a projector, until Bruno freezes as the owner narrates the official suspension of his authority to project films over the second half of the 1940s, through his former membership of the Nazi party. In the second sequence, an elderly female cinema owner responds to Bruno's question as to why she maintains her cinema's projectors but publicly projects no films, reflecting that cinemas, in her view, now endure such contamination, degradation and exploitation that she will not authorise projections until great films are made once again (in *Kings of the Road*, "great" films are characteristically those of Lang). Until that moment, the figure of the projectionist has to be an annulled one, but the suspended projectors themselves must always be maintained in imminent readiness to project those great films, if they emerge again.

Whenever it is publicly projected, the film-artist Bill Morrison's celebrated work *Decasia* (2002) forms an astonishing outburst of filmic contamination and degradation, but in the material form of pre-1950s decayed cellulose nitrate film-stock rather than the pornographic degradation which *Kings of the Road*'s disabused cinema-owner coruscatingly reflects upon. In numerous filmic archives, Morrison researched and digitally transferred films, dating from the early twentieth-century decades, of such decayed and deteriorated sequences that they would ordinarily no longer be projected (and even merited destruction, bearing the vital signs of cellulose nitrate film's capacity for combustions of such intensity that they evoke the many early twentieth-century disastrous cinematic conflagrations, since even submerged in water, nitrate's fierce fire will not burn out). In its projections, *Decasia* conjures corporeal revolutions – especially through a repeated sequence showing the revolving figure of an ecstatic Sufi dancer – that resonate with the incessant material revolutions of the celluloid reel before and after it passes through the projector's gate. Morrison's disintegrating corporeal figures combat their own filmic decay: a boxer relentlessly hammers his gloves against a seething eruption of celluloid contamination that envelops the entire middle section of the frame, supplanting the space originally held by the boxer's punchbag. Cities, too, have fallen into celluloid's engulfing decay in *Decasia*, projected now as negated wasteland spaces that oscillate between the just-recognisable forms of great urban avenues and their abrupt dissolution into scrambled, contaminated cities. The depth of projection's filmic decay (in *Decasia*'s amalgam of volatile nitrate-film sequences and the digital image, through Morrison's digitisation of those sequences) presents a mesmerising ocular spectacle. Its projectionist projects the abyss of film's fallen corporeal and urban histories.

This book, beginning with its accounts of Muybridge's itinerant 1891 Central European moving-image projection tour and of his 1893 construction at the Chicago Exposition of his Zoopraxographical Hall in which to project his sequences to spectators in a dedicated spatial location, explores the preoccupations and obsessions surrounding the figure of the moving-image projectionist. In particular, it focuses on projectionists who have themselves generated or created the moving-image sequences which they then go on to project, as with the work of Muybridge, the Skladanowsky brothers and Teiji Furuhashi.

Projection always formed a malleable profession with fluctuating dimensions, extending from projections to audiences of many millions at ephemeral World Exposition auditoria, to projections for the seemingly stable mass-audiences of static film-palaces and other twentieth-century auditoria, to projections in improvised venues for audiences of just a few obsessives in experimental art-film contexts. The projectionist may be itinerantly "on tour" or "for hire", constantly on the move but remaining preserved within an enveloping acute solitude that also surrounds the fixed-space projection-box of a cinema, or else the projectionist may be located within the heart of the auditorium itself, exposed to the audience and receiving their adulation or abuse, located corporeally among them or else standing directly beside the projection screen; conversely, the projectionist's figure may also be a hidden one, located behind that screen, mysterious and even devilish in that zone of invisibility. Through projection's incessant misfirings and integral malfunctions, the projectionist can also exist in negative interaction and confrontation with protesting audiences that have assembled to witness moving-image projection but experience instead a technological calamity or a revelation of the projectionist's ineptitude; then, the projectionist may exchange insults with the audience from the projection-box, or attempt to suspend or concertina time for the immediate interval required to rectify torn celluloid or an overheated projector. On very rare occasions – twice only, in Moldova and Cambodia – I have seen an enraged audience pierce or infiltrate the projection-box's door (as Bruno does in *Kings of the Road* in order to confront the masturbating projectionist), expel the projectionist from his domain, as though in an act of cinematic revolution, beating him if necessary and ejecting him from the cinema, before then attempting themselves to continue the projection.

This book focuses above all on the multiple experimentations of moving-image projection, from its glass-disc format of 1891 to the digital media of the contemporary moment. It explores experiments in the projection of the moving image – often interconnected with art and performance – of such momentous, unprecedented innovation that they became pivotal presences in determining the future of corporealities and urban space, across the twentieth century and into the first decades of the twenty-first century.

The chapter on Muybridge's 1891 Central European projections identifies the final phase of his innovative work as being that of an experimental projectionist, vitally preoccupied with creating spectators for the public projection of the moving-image sequences he had developed from his expansive work of the preceding decade or so, by travelling to those spectators' cities, and oscillating between projecting his work in scientific or technical institutes and in artists' clubs, always to awestruck audiences. After his 1870s-80s experiments in human and animal locomotion, and in panoramic, landscape, urban and war photography, Muybridge took on the status of the first moving-image projectionist. Muybridge always stood directly alongside his projection screen, vocally accentuating or narrating his projections. Several of Muybridge's auditoria still exist in contemporary Central European cities, still resonating with the furore of his corporeal and urban projections. Muybridge's handwritten correspondence from his 1891 tour emphasises its glory, alongside his own itinerant solitude as a projectionist.

The chapter on Muybridge's projections at the 1893 Chicago World's Columbian Exposition emphasises the reversal he then undertook in ceasing his tour (intended to continue to East Asia), and instead installing himself on the periphery of that Exposition's experimental metropolis. To do so, Muybridge devised his final unprecedented innovation as a simultaneously architectural, cinematic and corporeal one, by designing his Zoopraxographical Hall as an auditorium intended to assemble bodies whose eyes would be those of the first audiences for moving-image projection in a purpose-built space. Through the conjunction of the Exposition's pressurised urban dynamics and Muybridge's own acute obstinacy as a projectionist, his experiment misfired, its bodies emptied out and his building was soon infiltrated and occupied by another Exposition attraction. Although Muybridge's auditorium itself was soon razed and only subsists in its appearance in a scattering of photographs, other buildings in Chicago by its architect survive, along with Muybridge's colour projection-discs and other traces of that seminal projection experiment.

The chapter on the Skladanowsky brothers' corporeal and urban projections in Stettin and Berlin explores how exposure to the experimentation of Muybridge's moving-image projection-tour in Berlin in March 1891 led to his work's adaptation over subsequent years by young, touring magician-inventors who subsequently seized on the availability of newly-invented filmic celluloid from 1893 to transfer their own glass-disc projections to celluloid format, before undertaking their first public moving-image projections, mostly corporeal figures of dance, in Berlin in 1895. Crucially, the Skladanowsky brothers projected their work from behind the projection screen, enhancing the spectral strangeness of their projections of films which they had shot themselves (and which they projected, as with Muybridge's Zoopraxiscope, with a projector of their own construction, the "Bioskop"). The Skladanowsky brothers, drawing on Muybridge's urban preoccupations, finally transformed their work's focus from bodies to cities, extensively filming Berlin for the first time, and projecting those city-films only once, in Stettin in 1897, before they vanished from film's histories.

Muybridge's projections at the Chicago Exposition, in defiance of their 1893 aura of calamity, inspired future immense-scale World Exposition projection experiments as well as far smaller-scale experiments such as those of the Skladanowsky brothers. Pre-eminently, Muybridge's influence in initiating moving-image projection in purpose-built auditoria was tangible in 1970 at the Osaka World Exposition, which was devoted almost entirely to experimentation in immersive and sensorially focused moving-image projection, within pavilions designed exclusively to accentuate those projections. Archival films of the pavilions' interiors demonstrate the extreme sensory, corporeal and ocular demands which the environments placed upon spectators, as those audiences experienced films evoking corporeal and urban futures. The Osaka Exposition, with its 64,000,000 visitors, its unlimited financial and technological resources, and its contrary amalgam of corporate culture and avant-garde art, also demonstrates how the obsessions and aberrations integral to Muybridge's work – and more widely to the figure of the moving-image projectionist – can attain prominence in such a situation of experimentation and acute excess.

Muybridge's work, with its sequences' choreography of relentlessly walking, running, turning and falling naked figures, formed a primary inspiration for the moving-image projection work of the artist Teiji Furuhashi, *Lovers* (1994), which is often viewed as the first significant work in the then-emergent field of digital-art performance installation. Furuhashi's work, with its isolated, never-meeting corporeal figures inhabiting in darkness an immersive digital environment that surrounded its spectators, appears at first sight the antithesis to the Osaka Exposition's immense-scale corporate excess, though it too was funded by a technology corporation – Canon Inc. – and resonates with the financial and technological imperatives of Muybridge's own patronage, projections and projection-tours. In its envisioning of future corporealities via projection-focused experimentation, *Lovers* is also a work of profound obsolescence that intimates the acutely rapid transformations, integral to moving-image media, that had contributed to Muybridge's Chicago calamity. By 2016, *Lovers*' emergent, idiosyncratic mid-1990s digital technologies of projection had been rendered entirely non-functional, and had to be reworked from zero and standardised in order for *Lovers* to be projected once again for new art-museum audiences.

The book's final chapter examines how Muybridge's experiments, embodied in his European projection-tour and Chicago-auditorium projections of 1891-93, vitally prefigure contemporary projections' digital invocation of new corporealities and urban environments, emerging from the immense image-screens of worldwide cities as well as from the minuscule smartphone screens that guide bodies through cities, incessantly overlayering and reconfiguring urban space. The volatility and catastrophic malfunction integral to Muybridge's work also forms a pivotal element of those digitised environments, which may imminently disappear as utterly as the glass-disc and celluloid projections that preceded them. Even in its cinematic obsolescence, the figure of the moving-image projectionist, in its many manifestations – above all, that incorporated by Eadweard Muybridge – remains an evocative agent of visual cultures. The apparent irretrievable loss of the unceasingly awake, isolated and excessively concentrated film-projectionist's figure only accentuates still further that figure's status as a primary instigator of the contemporary world's relentless regimes of projection, and of its perpetual mutations and reinventions of corporealities and urban spaces.

Muybridge's Central European Projections, 1891

Across the Spring and Summer of 1891, Eadweard Muybridge toured the cities of Central Europe with his Zoopraxiscope moving-image projector and his hundreds of glass projection discs, projecting his moving-image sequences to awestruck and entranced audiences. He was then at the high-point of his notoriety, and artists, celebrities, scientists, politicians, anatomists, aristocrats, image-thieves and royalty packed into the venues in which he undertook his projections. Almost all of the responses to those projections – both corporeally and vocally, inside those auditoria, in the immediate response to Muybridge's work as a moving-image projectionist, and also in subsequent written accounts in the arts-journals and newspapers of the cities he passed through – were laudatory, to an extreme extent, and almost uniformly so, in those accounts, because they often simply replicated Muybridge's own adroit self-written and widely distributed press-releases. Muybridge moved from city to city, southwards, from Berlin – Prague, Vienna, Budapest, Munich, Rome, Naples, among many other cities – then back to Berlin. An autocratic and irascible figure with his distinctive long white beard, long and unkempt hair, and black suit, Muybridge was on tour, in glory, able to deliver for his spectators something which they had not seen before and which unfailingly provoked extraordinary sensations: moving images, primarily of bodies and cities. But, located at that precipice moment of Europe's immense ongoing urban technological upheaval and innovation, Muybridge's tour could hold together its intensive glory only for that instant, and no longer; his itinerance as the first moving-image projectionist possesses an aura of acute vertigo – imminently ready to pivot into calamity and disintegration – which would render his Central European tour fiercely unique but unrepeatable, and lead to its rapid historical oblivion.[1]

The figure of the transient moving-image projectionist – constantly assailed by a state of technological emergency in always-unfamiliar surroundings, and needing perpetually to improvise a way to keep going, through incessant obstacles and engulfing malfunction, in the movement from city to city – is one which fixed-location projectionists, in their sedentary projection-box domains, would view with contempt and derision across film's first decades and beyond, as though mobile projectionists were sub-standard amateurs or opportunists. Once cinemas with dedicated architectural forms began rapidly to proliferate in all cities of Europe and the USA, from around 1907-08 onwards, the moving-image projectionist occupied a clearly identifiable urban location; to be outside or excluded from the particular spatial regime of the cinema appeared outlandish and even criminal to many projectionists. Often, in the 1910s and 1920s, itinerant projectionists, unable to find employment in cinemas, or simply preferring itinerance, would move beyond urban space altogether and tour rural locales where no cinemas yet existed, projecting their films outdoors at the fall of darkness, either in village squares or on wastelands at a village's periphery. The writer and filmmaker Jean Genet's first infatuated contact of that era with moving images – in the village in France's isolated Morvan region to which he had been dispatched from Paris at birth as an unwanted bastard child – came through the intervention of itinerant projectionists who, supplanting the attractions of previous magic or magic-lantern shows, captivated their village audiences under starlight with incandescent moving-image sequences of dynamic, sexualised urban bodies. In the years immediately before and after the First World War, those open-air itinerant projections (still without synchronised sound, and accompanied by live music or the live voice of a commentator) possessed an immersive sensorial dimension of disjuncture and uprooting, often impelling spectators immediately to abandon their rural lives and head for the nearest megalopolis. Those projections served too to initiate or indoctrinate their spectators into how corporeality appeared to operate in urban environments: above all, through violence, deviance and ruthlessness.

Muybridge's Central European projection tour was entirely urban in its zigzagging itinerary of cities, and was undertaken at a time when no purpose-built auditoria for moving-image projection existed yet. As a result, Muybridge had to locate his projections within architectural environments that would enable experimentation, either in artistic or technological frameworks. His tour's status was unknown and uncertain, and its resulting volatility may have helped to attract the engagement of his audiences, who must have been aware that Muybridge's projections could not be replicated and belonged only to that instant. No precedent existed for such a tour, in its extremes and contradictions, which oscillated, among other aspects, between the collective rapture with which audiences received Muybridge's projections and his own figure of acute solitariness as their activator. Muybridge was a determinedly solitary man. Once he had irreparably destroyed his own San Francisco family-life in 1874, seventeen years earlier, by shooting dead his wife's lover (a murder which he admitted, but for which he was exonerated at his trial) and then consigning his sole child to an orphanage after his young wife's subsequent death, he no longer owned a house and pursued a solitary, frugal boarding-house or cheap-hotel existence, always inhabiting one small room and, as his occasional visitors remembered, engaging in his only indulgences of smoking innumerable inexpensive cigars and eating strong Sardinian cheeses. The expansive extravagance of his moving-image projections' inventiveness existed in contradiction with the sheer austerity of Muybridge's corporeal life. He pursued that cheap-hotel or boarding-house existence throughout his Central European tour, from city to city, from Berlin to Italy and back to Berlin across the months March to July 1891, as well as for the following decade. He would only own (or co-own, with family members) a house again at the very end of his life, in 1902-04, several years after his 1894 return to his birthplace,

Kingston-on-Thames, and after almost three decades of itinerance. Immediately, his irrepressible spatial obsessions materialised again in that "fixed" domestic environment, with his usage of his extensive back-garden's space to bury his photographic negatives, so deeply that most have never been rediscovered, and to begin singlehandedly to dig large garden pools designed to resemble the Mid-West USA's Great Lakes, the required arduous exertion apparently prompting his own death.

Muybridge's Central European tour was an extensive journey, from city to city, from railway station to station. Rather than restricting his demonstrations of his work to Europe's pivotal, prestigious locations for presenting innovations in visual arts and technologies, in Berlin and Vienna, he kept going, in obstinate itinerance, with the sense of sheer endurance that is invariably mediated by the figure of the moving-image projectionist. Muybridge appears to have booked all of his engagements himself, and often arranged several projections in the same city, even in venues located on the same central square, as in Vienna, where he projected his work in auditoria on either side of the Karlsplatz. He showed no interest whatsoever in the financial aspects of his tour. Instead, that sheer excessiveness of Muybridge's tour resonates with the many thousands of sequences he had photographed during his Philadelphia *Animal Locomotion* studio-experiments of the preceding years, when he had choreographed his often naked models to undertake almost identical gestural sequences, again and again, with only the distinction of a physical convulsion or technological malfunction to differentiate the models' movements. In his moving-image projects, Muybridge worked intimately with repetition and multiplicity, and his Central European tour manifests that compulsion of movement, which appeared also in his handwritten lists enumerating the names of the many cities he was traversing.

By touring from city to city, the itinerant projectionist actively transports moving-image sequences to audiences, rather than expecting those audiences themselves to travel to see them of their own volition. Whenever projectionists, as with Muybridge's work (and that of the Skladanowsky brothers, and of Teiji Furuhashi, in further chapters of this book), have previously filmed or recorded those sequences themselves, that compulsion to project has a special air of determination. The moving-image projectionist travels in a condition of solitude, for whatever distance or time is required, in order to reveal moving images to the eyes that wish to seize them; the projectionist is powered by obsession, and will rarely neglect, divert or annul that duty to project, which can be negated only by an overruling exterior factor, such as projection's obsolescence. Muybridge's extensive tour indicates the desire directly to engage audiences with his experiments, necessitating the corporeal amassing within an auditorium of spectators for exposure to moving-image projection. During Muybridge's Central European tour, that process of spectatorial amassing is undertaken only gradually and cumulatively, over the course of several months and across many adapted, pre-existent venues, through Muybridge's own restless mobility and via the alliance of incessant urban travel with his moving-image sequences of corporeality in movement. Two years later, in 1893, at the Chicago World's Columbian Exposition, Muybridge would abruptly reverse that working model, by designing and constructing the first auditorium solely designed for spectators to amass within in order to experience moving-image projection, and by himself staying put in Chicago in order to inhabit his own fixed-space auditorium. But as Muybridge's figure travelled incessantly from city to city through Central Europe in 1891, his projections, audiences and auditoria were all conjoined in dynamic spatial movement.

Once Muybridge had ascertained that his Zoopraxiscope projector was functioning perfectly, he would leave its operation in the hands of an assistant or surrogate projectionist, and walk directly through his audience, from the back of the darkened auditorium to a dedicated space alongside his moving-image screen and to its right-hand side, to give a vocal lecture-performance in which his own corporeality was intimately in conjunction with his projections; his screen was around twenty feet high and twenty feet wide. That vocal performance occupied exactly the same time as the moving-image projection. Muybridge's projections were usually undertaken without a break, and on his Central European tour he would speak malleably and unpredictably across time, from between an hour and a quarter to three hours, beyond linear time and against duration, driven vocally by his inspiration. Muybridge voiced moving-image projection as it manifested itself beside his body. His assistant had to watch that erratic body and the wooden pointer which Muybridge wielded in his hand, and to listen intently to his voice's urgency, in order to ascertain exactly when to accelerate or to slow the speed of the act of projection, or to switch from one set of projection discs to another. Muybridge's images for projection – adapted from his experimental multiple-camera work undertaken in Philadelphia and other sites – were imprinted sequentially around the peripheries of his discs, with those revolving images illuminated for moving-image projection by the Zoopraxiscope's lamp. Above all, Muybridge's assistant needed to know the pivotal instant when, having begun his lecture with discussions of his still images, Muybridge suddenly enthralled his audiences by transforming the still image into the revolutions of the moving image.

Muybridge was an immediately recognisable figure for his audiences, with his long hair, flowing white beard, and severe stare; he installed a large framed photograph of himself at the entrance to his projection venues or else in the foyer between the entrance doors. He accorded the unique stamp of his own voice's infinite authority as that of the revered initiator of moving-image sequences, naming himself "Professor" Muybridge (though the University of Pennsylvania, where he had undertaken his several years of *Animal Locomotion* experiments prior to his tour, never gave him any professorial title and treated him with ambivalence). His self-voiced exceptional status as an experimental technologist and artist allowed him at least an equivalent status to anyone present in his audience – the aristocrats or royalty, the generals or scientists, the artists or writers – or else enabled his status to far exceed that of his spectators, as with the upstart image-thieves and young touring magicians such as the Skladanowsky brothers who were at his Berlin projections to sieve his images and words in order to activate their own ambitions. Muybridge directly faced his audience, rapping his image-illuminated screen with his pointer to accentuate his words, as he vocally narrated his sequences of the actions or movements of human beings, in the forms of the athletes and models whose gestures he had choreographed at his open-air studio located in one of the university yards in Philadelphia, and of the animals, from horses to camels, monkeys to tigers, and birds such as crows, whose movements or flights he had recorded with multiple cameras in the Philadelphia zoo and other sites.

Muybridge was engaged in vocally analysing his moving-image sequences: piercing and infiltrating them, inhabiting their surfaces and the intervals between their images, traversing the space from the originating glass of his projected discs to their momentary corporeal inscription in light on his screen, and back again. He interspersed his sequences of human and animal movement with single slide images of artworks (often, from ancient Greek or Roman sources) that implied a historical precedence to his own sequences, while always emphasising a simultaneous negation of precedence: nobody had ever animated, activated or livened corporeality as he had done, through moving images. Occasionally, in the final part of his vocal lecture-performances, Muybridge also projected images of the cities in which he had worked, pre-eminently San Francisco, the urban space of which he had transmutated into the forms of immersive panoramas, always – as with his corporeal projections – fractured by an awry element of urban misfiring, since Muybridge invariably exposed his work's malfunctions as an integral component, alongside its repetitions and obsessions.

Muybridge's 1891 vocal lecture-performances – undertaken on tour across the cities of Central Europe, with his own body exposed to his audiences just as he exposed his screen-projected moving-image work with its incorporated malfunctions as well as its precisions – form the initiating manifestation of an intermittent performance-art strategy which eventually became a particular focus of worldwide experimentation in the 2010s, reflecting that era's proliferations of digital sequences. In contemporary lecture-performances, an artist, filmmaker or curator of moving-image sequences presents them live to an audience while being corporeally and vocally present in the auditorium, usually directly adjacent to the screen and, in that exposure, possessing a greater or lesser degree of vulnerability. Notably, the Lebanese performance artist Rabih Mroué undertook many such works across the 2010s, from *The Pixelated Revolution*, 2012-15, to *Ode to Joy*, 2016-20, in which he sat at a table with an opened MacBook, vocally interrogating for his audience the moving-image sequences he projected on a large screen (many of them presented as recent or contemporary found-footage, such as that of Syrian military snipers' killings of citizens attempting to film them with iPhones, or archival-historical footage of 1960s-70s Palestinian terrorist acts). In Mroué's work, such footage – often filmed by himself even when presented as "found" footage – demands the immediate presence of the artist's voice to incise and anatomise it, especially in its activations from stalled still images to death-inflected moving-image sequences.

While Muybridge's lecturing voice would have been always humanly "familiar" to some extent for his audiences, his moving-image sequences' projection was contrarily unprecedented, so that his vocal intimacy combined with the raw shock of moving-image projection must have launched volatile amalgams of sound and image in his spectators' perceptions. Muybridge lectured in English on his tour of Central Europe, whose cities were mostly German-speaking (though for some of his projections in the cities of Italy, he was able to speak in Italian); with the loud volume of his resonant voice, he was able effortlessly to reach the far peripheries of his auditoria, which usually held several hundred spectators. Sometimes, as with Muybridge's projections in Vienna, a translator was available, so that the intricate corporeal strata of Muybridge's own vocal interrogations of his moving-image sequences became intermedially echoed, in spectrality, through the figure of another, alien larynx, tongue and body, and were repeated, in near-synchronicity or else at an infinite distance from Muybridge's own idiosyncratic vocal analyses. At other venues where no translator's figure or voice was present to stand alongside Muybridge, and where many of his spectators certainly knew no English, his words must still have intersected in a strangely adhesive vocal rapport of revelation with his projected sequences, since his Central European audiences appear always to have been equally enthralled and awe-struck as they emerged from Muybridge's projection auditoria, whether they had understood his voice or not.

Alongside the vocal intermediaries who voiced into translation Muybridge's commentaries, he always had to depend also on all of his tours upon moving-image projection intermediaries. While the itinerant moving-image projectionist must exist in a state of acute solitude, the technological demands of live projection simultaneously exacted the splitting of Muybridge's status between that of projectionist and that of vocal lecturer. Muybridge had designed his Zoopraxiscope projector himself and commissioned its construction, incorporating a Singer sewing-machine handle to propel its movement. The projector constituted a unique, irreplicable artefact, non-standardised and impossible to mass-produce, in contradistinction to the industrially manufactured moving-image projectors that proliferated from the following decade onwards. Only the Skladanowsky brothers' equally idiosyncratic Bioskop celluloid moving-image projector of 1895 (which the brothers not only designed, but hand-built themselves) would possess that same aura of irreplicability. Muybridge only ever used one projector. He constantly updated and upgraded his Zoopraxiscope across the span of sixteen years from his first moving-image projections of 1881 to his last projections of 1897, renewing its malfunctioned or worn elements, adding new fans or other innovations, but never replacing his projector. It needed to be transported delicately on tour, from city to city, as an intimate corporeal accomplice or double to Muybridge's own body and voice. Before his projections began, it was always Muybridge who ascertained that the Zoopraxiscope was working well and was ready to project, but at a certain point in his performances, he had to leave it behind, in the care of a rapid-handed assistant, in order to traverse the auditorium and take his position of authority beside his projection screen.

Muybridge's intermediaries of projection are almost all mysterious figures who have fallen through moving-image projection history as anonymous assistants whose sole purpose was to carry through Muybridge's orders, since he could not position himself in two areas of his auditoria simultaneously. The Zoopraxiscope needed incessantly to be watched and fanned to prevent overheating, while Muybridge's discs had to be adeptly positioned for projection, at great speed. Almost no information on Muybridge's surrogate projectionists survives, and he appears to have employed new assistants for every tour, advertising in photographic journals or locating them through word of mouth, as though he were a touring musician offhandedly hiring a pick-up backing band for each tour. He needed young figures without commitments, since his tours extended over months of arduous travels. Only one of Muybridge's young projection assistants can be identified, Ernest Webster, since he wrote a letter to the librarian of the Kingston Museum in 1931, over forty years after his travels with Muybridge on the 1889-90 projection tour of the UK and Ireland which preceded the Central European tour: "I operated the Zoopraxiscope for him [Muybridge] over sixty times and we never had a breakdown... We had to travel all over England, also to Scotland and Ireland with much delicate apparatus and a quantity of compressed gas in tubes, about eight feet long. If the quick lime cylinders got damp... they would crack and crumble, perhaps melt the supporting pin and endanger the whole show but we always kept going somehow."[2]

Webster's memorial account of the 1889-90 tour must also directly evoke the experiences of Muybridge's assistant for his subsequent Central European tour, whose name (or names, since Muybridge may well have employed successive assistants as he travelled from city to city) is entirely eroded out of moving-image history, if it ever existed there at all. Webster's own name only hangs by a thread, since in all of the many press-accounts of Muybridge's 1889-90 tour, he is mentioned only once, as "Mr Webster, operator", for Muybridge's projection of 14 February 1890 at the Belfast Young Men's Christian Association Hall. Webster emphasises the relentlessness of their travels together, during which he and Muybridge very rarely spoke to one another. Often, Muybridge's projections would be such a success that he and Webster would travel on to their itinerary's next city before then doubling-back to the previous city to give a repeat performance. Muybridge's UK tour emanates nothing of the sustained glory and adulation he would go on to experience in Central Europe; the UK projection venues oscillated between the packed auditoria of monumental Victorian town-halls of the North of England and badly-attended, ill-judged venues, as with his projection at the Gloucester Corn Exchange on 8 December 1890 during which many rats scurried across the floor, driving Muybridge incandescent with anger.

Webster also evokes the caustic intolerance with which Muybridge responded to any flaw or deficit of speed in the delegated operation of the Zoopraxiscope (again, like a touring musician harshly castigating a hapless pick-up band) and which must also have been his characteristic response to any mishap during the subsequent Central European tour. A barriered distance between Muybridge and his "operator" is tangible in Webster's account. Muybridge also wrote in advance to the managers of all of the venues in which he undertook his projections, giving them exact specifications and demands which were necessary to prevent incendiary malfunctions of the gas-driven projections as well as to ensure that those projections seamlessly transmitted Muybridge's glorious technological authority and innovation. Webster's use of the word "show" for Muybridge's projections tilts his work into the domain of entertainment that was anathema for him, with his overriding obsessions with unprecedented experimentation in art and technology, though Muybridge was evidently engaged with his audiences, making them laugh whenever he could as well as inciting their awe.

The figure of the temporarily hired intermediary of moving-image projection – necessarily positioned between Muybridge and the corporeal immediacy of his work – accentuates the vulnerability and desired solitariness implicit in his own first embodiment of the figure of the touring moving-image projectionist. Delegating the duty of projection nightly, for the exact duration of Muybridge's lecture-performance, only renders more raw his own isolation beside his screen as he interwove his authorial voice with his moving-image sequences. Above all, Muybridge had to delegate to his assistant the pivotal moment of his projection's aura of spectacle, when the initial slide-projection of still images was abruptly transformed into the Zoopraxiscope's animation of moving-image sequences, with that revelatory instant of activation lost from Muybridge's own hands. His apparent disdain and forging of only-temporary, glacial rapports with his touring assistants indicates how deeply that loss of moving-image authority must have been experienced by Muybridge.

The moving-image projectionist, confined to the space of a projection-box (as with all cinema-projectionists from the end of the 1900s onwards), can never delegate the act of projection and is usually unseen in the performance of projection, whose operation only comes unstuck through technological misfirings, extreme instances of which may rip away the projectionist's habitual invisibility and expose that figure to being castigated or even ousted by enraged spectators. In cinematic projection, figures of authority may manifest themselves beside the screen to vocally explicate the action (and sometimes also to translate it), as with the figure of the *benshi* in Japan's silent-era cinema culture, but that commentator is never simultaneously a projectionist. Muybridge's experience on his Central European projection-tour forms an unprecedented but also outlandish one; he must be the activator and supervisor of projection, responsible for ensuring there will be no "breakdown", but at the same time is compelled to surrender that regulatory status to his subordinate's figure in order to stand exposed beside his projection screen, deploying his voice and corporeality to captivate and enthral his audiences with his commentaries on his own moving-image sequences. His body is aberrantly split in that act: static in its vocal positioning beside his screen but also incessantly thrown across the auditorium's space from the originating site of projection, as with his sequences' own propulsions in light.

Muybridge always intended his experiments in moving-image projection to be thoroughly documented for future projectionists and researchers. From the 1870s onwards, he amassed documentation and press-accounts of his work, transporting that material with him on his travels in the USA and Europe with the evident intention that, at some point, he would create an immense book of projection (alongside the other dimensions of his work, especially in photography) that corresponded to his corporeal and vocal experimentations. Almost all of the final decade of Muybridge's life, from 1894 to 1904, would be devoted to the generation of that future-oriented book, or "scrapbook", which he then bequeathed to the Kingston Museum along with his projection discs and Zoopraxiscope projector. In his preoccupation with the future memorial status of his projection experiments, Muybridge's work is that of an anomalous projectionist, since projection is intrinsically ephemeral, burning up as it is enacted to its dedicated spectators who have assembled only for that act's momentary conjuration of moving images. The obsessional solitariness of moving-

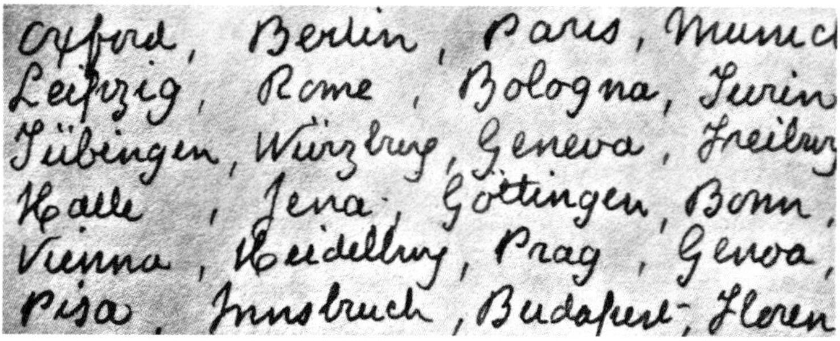
Oxford, Berlin, Paris, Munc
Leipzig, Rome, Bologna, Turin
Tübingen, Würzburg, Geneva, Freibur
Halle, Jena, Göttingen, Bonn,
Vienna, Heidelberg, Prag, Genoa,
Pisa, Innsbruch, Budapest, Floren

Ill. 3: Eadweard Muybridge, letter fragment, 15 July 1891.

image projectionists entails their subjection to oblivion as soon as they vanish from their projection-box, terminally exhausted or wiped-out by their addictions; cinema projectionists in cities worldwide often became notorious for their sheer endurance (able to occupy their projection-boxes without respite for months or years on end), and for their technical adroitness or ineptitude, but once they became evicted or voided in one way or another from their site of projection, they were invariably forgotten. The transient moving-image projectionist too is a figure without a future – as with Muybridge's "operators", nameless and near-lost to all histories of projection – who risks disappearing without trace in transit between cities, as with one of Muybridge's competitors, Louis Le Prince, who vanished during a train journey between Dijon and Paris while developing a three-lens projection device (which he never had the opportunity to demonstrate in public), six months before Muybridge's Central European projection tour began.

Muybridge intended his work as a moving-image projectionist to be comprehensively recorded, primarily by himself. On the completion of his Central European tour, after he had returned to its starting-point in Berlin, he wrote a long letter on 15 July 1891 to Jesse Burk, a young administrator at the University of Pennsylvania whose work involved helping Muybridge to promote and distribute the *Animal Locomotion* publication based on Muybridge's motion-photography work of the preceding years in Philadelphia. Muybridge's letter exists to embody his Central European tour. In its first lines, he announces: "I will give you a brief account of my happenings from the time I left Berlin in the Spring."[3] Muybridge then documents his itinerary southwards from Berlin, enumerating a number of the venues in which he projected his work, but sometimes only invoking the many cities he traversed, in insistent sequences of those cities' names. Often, his concern is to directly link the venues or institutions in which he projected his work to their subsequent acquisition of a copy of *Animal Locomotion*. The publication's great expense meant that it was usually bought institutionally, by artists' societies or scientific associations, for their libraries, but Muybridge also meticulously lists the wealthy or celebrated individuals (in Berlin, Hermann von Helmholtz and Rudolf Virchow, prominent scientific pioneers of eye-movement and anatomy, among other domains) who bought copies for themselves. In such instances, the ephemeral moving-image spectator of Muybridge's projections then also becomes the permanent proprietor of his *Animal Locomotion* publication.

Muybridge's letter to Burk underplays the near-total astonished adulation which his Central European projections had generated in their urban audiences, instead focusing on vaunting his own linguistic capacities – "As, of course, a man cannot travel without picking up some knowledge of the language I am happy to say that I now speak Italian, German, and French with equal fluency" – despite the press-accounts of his events being explicit that he habitually lectured alongside his projection screen in English.[4] Muybridge's letter oscillates between urban evocation in envisioning the sheer relentless excess of cities in which he projected his work (he occasionally jokes about having taken only-momentary holidays from the professional "business" of those projections), and the unprecedented experience of his instigation of moving-image projection together with the ecstatic response to it.

Muybridge had assembled cuttings of the press-accounts of his Central European projection tour, and sent them on to

Burk for safekeeping as future raw-material for his planned memorial book. Fragments from those press-accounts would subsequently be included in the promotional programme that Muybridge produced to accompany his 1893 projections at the Chicago World's Columbian Exposition, but either he or Burk appear to have somehow lost those cuttings between the end of Muybridge's Chicago projections and his return to the UK in the following year, since they were never incorporated into Muybridge's scrapbook when he finally assembled it in Kingston-on-Thames. Although all of those press-accounts can be reconstituted from national library and archival sources in Vienna and other cities, Muybridge's own personal documentation of his Central European projections became lost knowledge that perversely slipped through his adept fingers (those misplaced accounts of his tour constituted the sole missing material in the scrapbook which he assembled in Kingston-on-Thames to embody for the future his work's two decades of glory). The acute intensity of adulation that is pervasive in press-accounts of Muybridge's Central European tour, as well as his loss of that adulation's traces, intimate the correspondingly acute fragility of his work as a seminal moving-image projectionist.

Part of the adulatory response to Muybridge's Central European tour came about because he composed it himself. For the past decade, since the early 1880s, Muybridge had become highly skilled in generating a controlled media environment for his work. He assiduously produced detailed press-releases and sent them to every newspaper and arts-journal in the cities he travelled through, often having his press-releases translated into German and other languages. Since many journalists simply transcribed Muybridge's own advanced sense of the uniqueness and significance of his moving-image projection work, the accounts of it possess a degree of unremitting repetition (often recurring word for word) which also accords with Muybridge's own engagement in his work with repetition as a creative strategy. But at the same time, whenever journalists chose to set aside Muybridge's press-releases and preferred to write their own accounts – giving details, for example, of the packed projection-auditoria and the volatile aura of excitement generated by Muybridge's projections – that tone of adulation remains consistently present, as though Muybridge's visual sequences, projected from city to city and from auditorium to auditorium, were able to generate their own, autonomous written and spoken language, even if it existed in its distinctive homogeneity solely for the tour's duration.

The extraordinary spectatorial response to Muybridge's 1891 projection events is tangible in his handwritten letter to Burk, as well as in the press-accounts of those events, and is engrained too, sonically and visually, within the surviving auditoria in which Muybridge demonstrated his work. As moving-image performance-lectures, those events (as with all performances) flared and then evanesced within their duration, that vanishing exacerbated by the heated intensity of the response to them, in defiance of Muybridge's own desire that they (or traces of them) should endure into the future, to thereby demonstrate the innovativeness of his work as a moving-image projectionist, located in its many urban environments. Muybridge's Central European projection tour also formed an urban intervention that drew upon and also transformed the insurgent vision and dynamic corporeality of cities in their acute technological, ocular and auditory upheavals of the nineteenth century's final decade.

Those upheavals were most tangible in 1891 Berlin, where Muybridge began his Central European projection-tour with two events, on 9 and 11 March, at the Urania Society's lecture theatre, located alongside an astronomical telescope and laboratory, at the north-western corner of a large exposition park designed to highlight innovations in technology. Berlin was a uniquely experimental city in the 1890s, exceptional in its citizens' engulfing desire to witness and experience unprecedented cultural and urban manifestations, across art, architecture, research into corporeality and anatomy, sexual behaviour, technology, speed, transportation, among many other areas.

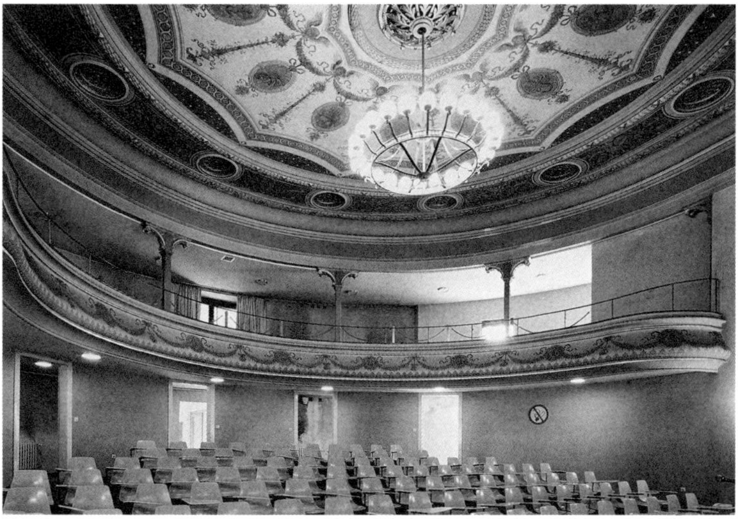

Ill. 4: Muybridge's Berlin Urania auditorium, now an abandoned auditorium of the Moabit police compound, 1, 2020.

That receptiveness to innovation rendered Berlin the ideal environment of that moment for Muybridge's moving-image projections; he presented his work there first, and also based the organisation of his tour in Berlin, finally returning to the city in July to assess the tour's overwhelming success, after his lengthy projection journey across Central Europe's cities. Muybridge's projections in Berlin also exposed his moving-image experimentation to intimate scrutiny, both by young image-thief aspirant projectionists such as the Skladanowsky brothers and also by tenacious competitors such as Ottomar Anschütz, who intended to elevate their own work against that of Muybridge, while sieving it for their future projections.

Berlin's location at the epicentre of immense transportation networks (especially train networks, alongside canals and rivers) meant that immense crowds had flooded into it especially over the past twenty years. New tenement districts, such as Prenzlauerberg, had been rapidly constructed to accommodate, in dense corporeal amalgams, several million menial workers arriving from Berlin's surroundings, especially from the regions to the city's east, to work especially in chemical industries and breweries. Visitors from other cities used the new transportation routes to converge upon Berlin in order to experience its self-defined status as an expansive *Weltstadt* (or "World-City") that embodied accelerating global innovation, in technology and art, as well as in spectacles of sexual experimentation. Grand hotels with lavish ballrooms equipped with many performance stages, such as the Central Hotel directly alongside the Friedrichstrasse railway station, had been constructed to accommodate those sensation-avid visitors, though Muybridge himself maintained his usual cheap-hotel regime in Berlin.

Muybridge's moving-image projections in Berlin attracted an audience engaged primarily with his work's incisions into corporeality and its movements and gestures. The capacity of the moving image to allow the spectator's eye, across a span of time, to map and enter into gesture – with its distinctive accelerations, malfunctions and repetitions – provided an unprecedented revelation. Time in its conjunction with corporeality was rendered multiple, sequential, able to reverse itself or else proliferate. Muybridge's audience was fluidly composed of scientists and anatomists, artists and photographers, as well as spectators exhilarated by the naked or transparently clothed bodies in his work. The many press-accounts enumerate the names of the enthralled scientists and artists in the audience – Hermann von Helmholtz, Ernst Werner von Siemens, Wilhelm Julius Förster; Adolph Menzel, Ludwig Knaus, Karl Begas – alongside prominent military figures such as Helmuth von Moltke (who died in the following month) attending the event specifically to witness Muybridge's sequences of horses in motion. Those accounts also highlight the audience's recognition of a sense of profound rupture in space and time, propelled and incorporated by Muybridge's projections. Muybridge himself, in his selection of press-responses to his Berlin projections for inclusion in the programme he prepared for his Chicago Exposition projections two years later, particularly emphasises responses to the vocal dimensions of his lecture-performance, highlighting his own auditory and corporeal presence alongside his projection screen: "The lectures of Mr. Muybridge are unquestionably the most intensely interesting we ever listened to."[5] Essential to Muybridge's extreme challenge to his Berlin audiences was the imperative for them to oscillate sensorially and in instantaneity between his moving-image projections (with their unprecedented ocular demands) and his commentating voice; they listened to his voice as it anatomised the moving images projected before their eyes. Many press-accounts invoke the sheer volume of the outbursts of cries and applause erupting from Muybridge's spectators throughout his projections, adding another stratum of noise or cacophony to its vocal dimension.

The semi-circular auditorium of the Urania Society (named after Urania, the goddess or muse of astronomy) had been opened only two years earlier, in 1889, as an initiative of Berlin's prominent scientists and astronomers, such as Förster, with the aim to create a public forum for audiences' direct contact with technological innovation. The venue's events programme was closely associated during the era of Muybridge's projections with the Urania building's adjacent astronomical telescope and laboratory; during the four decades after the Urania Society had commissioned a new lecture hall and moved away to it in 1896, the auditorium of Muybridge's projections became used primarily for astronomical lectures. Muybridge projected his sequences below an ornate painted circular ceiling representing the zodiac. The main ground-floor auditorium seated around 200 spectators, but the single balcony area above that auditorium had wide, extensive standing-room at its rear, enabling a maximum capacity totalling around 400 spectators (press-accounts of Muybridge's body-packed projections indicate that maximum capacity was exceeded, and many prospective spectators were denied entry). Muybridge had expected to project his work only once, on 9 March, but the exceptional response to the event led the Urania directors to insist he return two evenings later for a second event, for an audience of those excluded from the first projections alongside spectators attracted by ecstatic newspaper press-accounts that appeared on 10 March. Although Muybridge had a translator for his first projections on 9 March, no translator was engaged for the second event and he lectured directly to his audience. The response to his first two projections on 9 and 11 March was so awed and exceptional that he was invited back to the Urania to demonstrate his work at two further, well-attended events, on 16 and 17 March, before leaving Berlin for his moving-image tour's next projections.

As with all of Muybridge's projections during his Central European tour, a seminal instant for his spectators emerged when, after beginning with the showing of still-image slides, Muybridge gestured to his Zoopraxiscope's "operator", positioned directly behind the audience, and the darkened auditorium was suddenly transformed into a moving-image arena. That moment of entrancement generated collective vocal astonishment as Muybridge's sequences of human and animal locomotion were projected. But at least two spectators were dissatisfied: the projections were viewed with a caustic professional eye by Muybridge's competitor Anschütz, and the eminent art and photographic historian Bruno Meyer (who attended the second evening) attacked Muybridge in the specialist photographic-newspaper *Deutsche Photographen-Zeitung* for not advancing his work beyond his first public projections of 1881, though Muybridge's Berlin projections were drawn from his more recent *Animal Locomotion* project undertaken in 1884-87 in Philadelphia, and Muybridge's work is invariably possessed by an intractable obsessional repetition that negates any conception of its own internal variance or progressive advancement. That core of repetition, generated by Muybridge's innumerable sequential excavations into corporeality across several decades of sustained photographic and moving-image experimentation, exists in tension with the unprecedented, volatile status of his one-off 1891 tour; his audience in Berlin would possess only the unique opportunity of that tour to witness the live performance

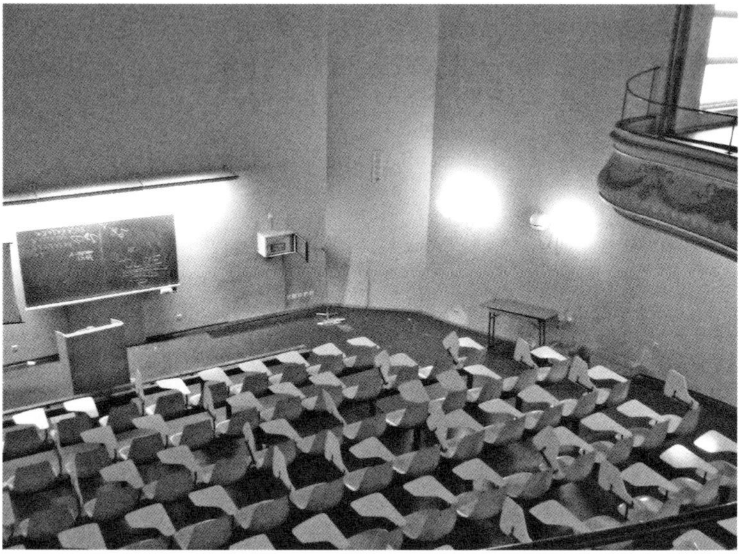

Ill. 5: Muybridge's Berlin Urania auditorium, 2, 2020.

of his projections. Embodying the figure of the transient moving-image projectionist who can never remain in stasis in urban space, Muybridge prepared to travel on from Berlin with his Zoopraxiscope.

Muybridge's Berlin projection auditorium miraculously survived the 130 years between his corporeal and moving-image presence there and the current moment, so that his projections possess an immediate spatial aperture into the contemporary city. On a cold February day, I left the southern exit of Berlin's new central railway station (built on the ruins of a previous railway station on that same site, the immense and majestic Lehrter Bahnhof, now vanished without a trace), and walked through a zone of shoddy concrete-block hotels to the fragment still surviving from the 1880s exposition park of exhibition halls, designed to demonstrate new technologies and scientific innovations, known as the *Universum Landes-Ausstellungs-Park* (or "ULAP"). Muybridge evidently prized exposition parks highly as ideal locations for his projections, as he would with the World's

Columbian Exposition park in Chicago two years later, and this appears to be why he proposed the public demonstration of his moving-image experiments – as an amalgam of scientific innovation and unprecedented artwork – to the Urania Society, which responded positively and eagerly invited Muybridge to show his work there. Alongside the elevated city-railway which horizontally traverses Berlin, a triangular strip of that "ULAP" park had been preserved, its moss-covered earth holding strips of benches, deserted and unused, but constellated with century-old trees and with the exposition park's original entrance stairway descending from the Alt Moabit avenue. The semi-disintegrated stone stairs zigzagged downwards, jarred and crooked as though subjected to a long-distant seismic uprising, half-heartedly barriered away from the park's non-existent visitors. In its depopulated silence at the heart of the city, under the winter sky, that park emanated its own askew, death-striated lineage: at numerous points of urban upheaval and breakdown in Berlin's twentieth-century history, extending from 1919's Spartacist Uprising to the USSR army's invasion at the end of the Second World War, the park had been used as an execution site. In 1934, at the beginning of the Nazi era, the park had been reconceived as a military aviation theme-park and display complex, its previous emphasis on technological innovation rendered redundant. Following the Second World War, its entire terrain became part of the interzonal wasteland dividing West Berlin and East Berlin, located on the Berlin Wall's western side following its construction in 1961.

I passed underneath the elevated course of the city-railway, and the Moabit-district police compound soon came into view at what had been the north-western apex of the exposition park, with twenty or so concrete buildings enclosed within barbed-wire topped fencing. Once I had turned the corner into Invalidenstrasse, the exterior of the semi-circular auditorium appeared, close to the compound's main gate with its signboard "*Polizei Direktion 3: Kriminalitätsbekämpfung*" alongside numerous warning signs deterring entry. Muybridge's auditorium had a strangely stranded urban history once the Urania Society had left it behind in 1896, five years after his projections; it was preserved more through oblivious neglect than by intention. After the four decades in which the auditorium was used for astronomical lectures, it was incorporated from the mid-1930s into the Nazi aviation park as a demonstration hall. The astronomical telescope and laboratory alongside it were destroyed – along with almost everything in the former exposition park – in wartime aerial bombing of 1943-45 and in the heavy street-fighting of the USSR's ground invasion of April 1945, and Muybridge's auditorium was heavily damaged but remained intact. It possessed a resilient iron internal framework, constructed to last during the 1880s era when Berlin's far-sighted architects had the motto: "the most expensive is finally the cheapest". The auditorium stood abandoned in the near-uninhabited postwar interzonal wasteland, its seating areas removed and used for the improvised storage of coal and wood, until the construction in 1963-65 of the Moabit police compound, which surrounded and enclosed it. The auditorium was then renovated, strengthened and electrified, in 1968, its internal walls plastered and the entire construction enclosed in a carapace of ochre brickwork. For several decades, it was used intermittently as a lecture-hall for police cadets, plastic chairs scattered around the auditorium and a blackboard positioned at the location of Muybridge's projection screen. Then, it was abandoned again.

I met *Kriminaloberkommisar* Mario Gleissberg outside the police compound's main gate. He told me he was based at another Moabit district police-station and had never heard of the auditorium before, and was curious to see it. I followed his shaven-headed figure in his long black coat as he gestured to the caretaker, Herr Stock, who unlocked the compound's gate and then took us to the auditorium's door. After he had unlocked it, it still wouldn't open, and he shouldered the door ajar, then wrenched it, muttering that nobody had been inside for a decade or more. I inhaled the thickly dust-laden atmosphere of the entrance corridor, then turned leftwards to the auditorium's entrance. I entered Muybridge's projection-auditorium from the position in which his Zoopraxiscope had transmitted moving-image sequences to Berlin's spectators for the first time, once Muybridge had gestured to his anonymous assistant to activate those sequences. The caretaker turned on the unblown bulbs of the ceiling's golden chandelier, and a few bars of sunlight entered the auditorium from windows around its balcony level, but it remained darkened. Directly above, the astronomical zodiac-inscribed ceiling was intact, its paint's design cracked but clear, and enclosed on one side by the semi-circular balcony with its ornate facings. That auditorium formed a profoundly spectral environment of the origins and end of the moving image.

Making my way down the ground-level sloped rake, between the strewn and upended plastic chairs, I reached the location of Muybridge's screen, alongside which he had stood, for four nights in March 1891, to vocally accentuate his projections. The police-cadets' blackboard was still there, inscribed with the numbered imperatives of criminal detection and, on one side, the chalked and underlined word "*POLIZEI*" in huge letters, as though those young cadets had once somehow become disorientated to the point of beguiling hallucination by the enduring atmosphere of Muybridge's moving-image projections, and had needed a sharp return to their duties of the immediate moment. Moving images themselves constitute a criminal preoccupation and obsession, their insurgency and denunciations of state-power often persecuted and rendered illegal, across their history; filmmaking is often a criminal's aberrant profession, as with Genet's film *Un Chant d'Amour*. I finally climbed the iron stairway at the rear of the auditorium and joined Commissar Gleissberg on the balcony, where he was gazing down into the auditorium. The balcony's now-emptied seating area was backed by the wide space where those of Muybridge's crammed-together spectators who were unable to secure seats had stood, glimpsing his sequences over other spectators' shoulders in the darkened auditorium's awestruck cacophony. The balcony's rear wall appeared vertically streaked with large patches of damp, holding the dimensions of human bodies, as though still marked by the exhilarated body-heat emitted by Muybridge's seething spectators. From the balcony, the zodiac-inscribed ceiling was intimately close, and it was possible to stretch an arm out over the iron balcony-railings and almost touch the desiccating-paint Taurus symbol of the month of Muybridge's death.

When it was time to leave, the caretaker lovingly pointed out the floor-level iron ventilation grilles that ran along the entire semi-circular back wall. Looking down through them, it was clear that the building's indestructible iron and brick framework descended deeply into the ground below the auditorium. The Commissar joked that, if we pulled open one of the grilles, we could probably find some of Muybridge's lost projection discs down there. From what his bosses had told him, there were no future plans for this auditorium, nothing at all; it would simply remain exactly as it was now, in suspension. I told him it was probably best to leave it that way, undisturbed and abandoned, inhabited only by moving-image projection's ghosts; but, in addition to his Berlin auditorium, Muybridge's Zoopraxiscope projector had survived too, along with many of his projection-discs, since he had returned to England with them in 1894 and eventually either bequeathed them to his local museum or else buried them in his back garden, so it would be conceivable, at some future moment, in this space, to reignite his projections, for new eyes. The auditorium appeared still incandescently charged with Muybridge's projections, from those nights 130 years ago, as though that auditorium's passage through the intervening time – via re-usage, obsolescence, violence and warfare, abandonment and interzonal stranding, and inhabitation by criminality's preoccupations – had been abruptly elided, generating a mysterious simultaneity between Muybridge's projections and the lived, immediate instant. The visual and aural residues of Muybridge's projections remained inextricably engrained in that auditorium's space, within the far perimeters of its ornate zodiac-ceiling, as well as enduringly located in the urban space of the city outside, with its infinite animated digital screens. Muybridge's corporeality, and his status as a transient moving-image projectionist, was densely engrained in that space, too. A few moments later, those obsessions of projection were suddenly overturned, the caretaker had re-sealed that lost, seminal space, Commissar Gleissberg had gone to catch criminals, and I was out on the streets of Berlin, at the heart of the wasteland, on the far side of the police compound's barbed-wire fencing.

Alongside those of Muybridge's Central European tour projections that can be inhabited tangibly and corporeally, as with those undertaken for ecstatic public audiences at Berlin's Urania auditorium, other projections' locations and exact dates are often unknown, whenever a projection has slipped through its archival records (or where the archives themselves have combusted, taking all trace of the projection with them). At the same time, in those lost cities of Muybridge's 1891 projections, it appears certain that he activated his Zoopraxiscope and revealed moving-image projection for the first time to his spectators – sometimes consisting only of a small, closed group of artists or an exposition's technologists – since he enumerates those cities in his tour's itinerary and in his lists of sales of his *Animal Locomotion* publication. In such cases, phantom projections materialise, in plausible urban locations conjured by Muybridge's few fragmentary indications. But however deeply those cities and their archives may be explored, they will not yield the material evidence of projection, which remains a spectral and mysterious act. Before leaving Berlin and heading south, Muybridge gave a further moving-image projection – this time a private, closed one – at what he calls in his scrapbook and Chicago programme the Berlin "Society of Artists" (*Künstlerhaus*). At that time, Berlin's Society of Artists had no fixed location or dedicated auditorium – it would acquire one only in 1898, in an ornate building close to the location of the Urania complex, seven years after Muybridge had left Berlin – and the society's events habitually took place in an improvised way at one or other of its members' studios. Muybridge was used to projecting his work in the confines of artists' studios, as he had done several times at the studio of Ernest Meissonier in Paris for that city's artists. Although Muybridge invariably stipulated his demanded conditions of projection in advance to venue managers, as his "operator" Ernest Webster emphasises, he was evidently ready to project anywhere, and could, for example, set up the Zoopraxiscope in the corridor outside an auditorium if he was dissatisfied with the venue's conditions of projection, as he had in Newcastle in September of the previous year.

From Berlin, Muybridge travelled on to Munich, where he gave a hugely successful projection-lecture on 16 April at one of the city's pre-eminent cultural venues, the Odeon, whose auditorium had been designed in the 1820s as a concert hall; its transformation into a projection venue formed a unique event. The building's exterior was constructed to resemble a palace, on a square of equally sumptuous buildings. The Odeon – viewed in that 1890s era as one of the most splendid, extravagant auditoria in Europe, with its frescoed ceiling – had a far greater capacity than the Urania, holding around 1,500 spectators in adjoining seated and standing areas. The curved walls of the stage area, at the southern end of the auditorium, held immense plaster laurels and granite columns that began at the building's mid-height. Columns also surrounded the oblong-shaped main auditorium. The building's colonnaded interior may have formed an inspiration for the columns Muybridge incorporated into his design of the following year for his own Zoopraxographical Hall to be constructed in Chicago for his moving-image projections at the World's Columbian Exposition. The Odeon's auditorium was largely destroyed by aerial bombing in 1944, but many of its columns, along with the building's resilient exterior walls that enclosed it, survived intact. In its subsequent 1950s reconstruction, the Odeon building was adapted for administrative use, losing its status as a concert hall. The auditorium's space in which Muybridge had projected his work was redeployed as an interior courtyard, with the even-

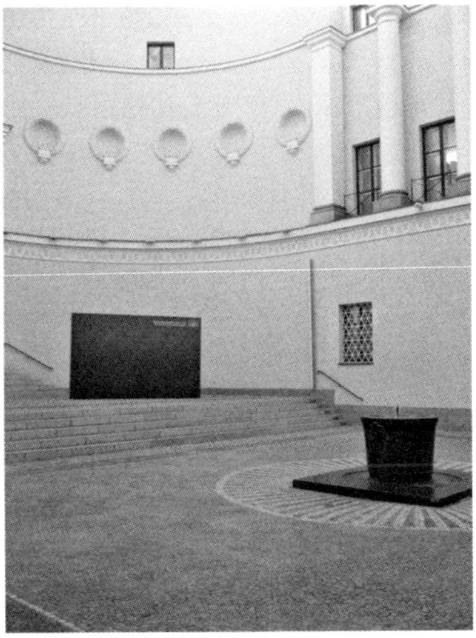

Ill. 6: Muybridge's Munich Odeon auditorium, 2020.

tual addition of a transparent roof, as though all memory and trace of Muybridge's tumultuous evening of projections there had been excoriated and voided, into oblivion. The reconstruction occupied exactly the original dimensions of the concert hall, including its stage area.

Despite the auditorium's size, Muybridge's Odeon projection event had been filled to capacity, and beyond, leaving many hundreds of spectators unable to enter. The Bavarian newspapers and arts journals, in their accounts of Muybridge's projections published on the following day, 17 April, emphasised above all how packed the event was, with its immensely enthusiastic and vocal audience. Those accounts also especially evoke Muybridge's projections of his sequences of women dancing, and of a naked female figure abruptly soaked with water from a wooden bucket, which he had developed from his Philadelphia work, along with evocations of Muybridge's then-current experiments in creating moving-image sequences of birds and insects. Muybridge himself singled out a Munich press-account from the prominent *Münchner Neueste Nachrichten* newspaper and included it, in translated form, in his programme for his 1893 Chicago projections as well as in the scrapbook which he worked on over the final decade of his life. That audience-focused account, with its emphasis on the event's conjoined aesthetic and technological strata, evidently seized the corporeal dimension of the responses to his work which most directly engaged Muybridge: "The lecture was given in a popular manner, with scientific accuracy and artistic taste... The room was filled to the last corner. Nearly all the [Bavarian] Royal Family and the Ministers were present."[6] (The current Bavarian King, Otto I, was mad, but still attended Muybridge's projections.) Muybridge's vocal commentary on his projections was translated simultaneously by a celebrated Munich-based artist, Robert Koehler, whose English was especially fluent since he had lived for much of his life in the USA. That Odeon event, with its astonished audience's extreme acclaim for Muybridge's work within a palatial location, must have appeared to him as one of the pinnacles of his European tour. Muybridge gave two further projection events in Munich – an event open to the public at a lecture hall (destroyed without trace during 1940s bombings) at the Technical High School (*Technische Hochschule*), another palatial building patronised by the Bavarian royal family, and a smaller-scale closed event intended solely for artists, held within an artist's studio – before moving on.

I arrived at 7am at the Bavarian Ministry of the Interior, Sport and Integration, after walking through Munich's still-darkened avenues, to be shown Muybridge's Odeon auditorium within it. My visit had to elapse before the ministry employees arrived to begin their work for the day. The guard met me at the doorway and shook my hand gravely, then led me along a wide passageway to the original brass-framed doors of the former Odeon concert hall. As he unlocked them, he told me he had to remain with me at all times, and watch for any acts that intimated terrorism, criminality or spatial outrage, warning me: "This is a Security Zone." The day before, I had watched Tarkovsky's *Stalker*, in which, at the end of the Stalker's journey through the Zone, his daughter, Monkey, telepathically projects glasses across a table, so I was already receptively immersed in zonal obsessions. As the guard unlocked the auditorium's doors, I asked him if Muybridge's projection "zone" contained a Sphere able to grant all desires, as in *Stalker*. The guard frowned at the film's name, and told me he far preferred Fassbinder. If, when I left here, I walked past the *Haus der Kunst* and across the shallow-flowing river Isar, I would reach a churchyard in which Fassbinder was buried under a tree with tiny red leaves. No, no moving-image projections were ever held in this auditorium now, only occasional receptions for ministerial bureaucrats. On the podium at the curved end of the auditorium where Muybridge's projection screen had been positioned, reached by five stone stairs, a freestanding digital screen was now installed where Muybridge had stood, alongside his own screen, to deliver his vocal commentary. The dawn sunlight was flooding through the transparent roof far above, illuminating the colonnaded auditorium that had been animated for two hours on 16 April 1891 with an incessant cacophony of awe and excitement, saturated with bodies, on the night of Muybridge's projection there. Met with adulation, it must have been his greatest projection. The guard returned to the doors, his boots echoing on the stone floor, and then there was silence, only the ineradicable residues surviving, via destruction and incineration, of Muybridge's projection. Finally, he returned: the ministry workers were arriving, I had to leave, and he repeated, as though he had not already said it: "This is a Security Zone."

Muybridge's projection event in Prague, located between his Munich projections of April 1891 and those in Vienna at the beginning of June, forms a phantom projection that can only finally be interrogated by conjectural topographical exploration, in the absence of any trace of Muybridge's presence in the city in its archives (many of them incinerated during warfare or now missing, after 130 years). Although Muybridge lists Prague in his tour's urban itinerary and as the location of a sale of *Animal*

Locomotion, I began to wonder after several weeks of investigation there if Muybridge had actually invented or fabricated his time in the city, in the same way that, on his return to Kingston-on-Thames from the USA, three years later, he asserted to members of his family that he had actually returned from Japan – a plausible point of departure, since the mid-1890s formed the era of Japan's newly emergent interest in photographic experimentation and of its presence as a new technology at expositions, such as the Fourth National Industrial Exposition held in Kyoto – thereby potentially creating an intricate urban bluff or obfuscation to render invisible his negative experiences at his actual point of departure, Chicago. I wondered too if Muybridge had enumerated Prague in his projection lists as part of a great urban delirium of naming that necessitated the expansion of his tour to unverifiable but conceivable projections in every major Central European city.

Finally, I walked out through the north of the city – over the Letná ridge still strewn with fragments of the dynamited debris from the 1962 detonation of Stalin's colossal statue that had stood there overlooking central Prague – to the 1891 Palace of Industry (*Průmyslový palác*), built for the Prague World Exposition or "Centennial Exposition" (marking a century since Prague's first industrial exposition) which opened on 15 May 1891, around the moment when Muybridge would have visited Prague between his documented projection events in Munich and Vienna. Through its amalgam of innovative technological display and crass fairground entertainment, the Prague Exposition had formed a low-key forerunner to the Chicago World's Columbian Exposition, with its grandiose official buildings and disreputable periphery, where Muybridge projected his work two years later. I started to convince myself – with zero archival justification, and as though irresistibly propelled by the obsessional dimensions of transient moving-image projection – that Muybridge must have undertaken his Prague projection there, perhaps as a low-key demonstration for industrialists or entrepreneurs who could conceivably sponsor his planned future work into the moving-image capture of birds and insects in flight, or else even as an opportunistic one-off sideshow to provoke momentary spectatorial awe in the Exposition grounds' margins.

As with Muybridge's projection auditorium in Berlin from the same era, the Palace of Industry had miraculously survived, the date "1891" inscribed prominently in immense iron letters on one side of its central hall. But the palace was now dilapidated and part-abandoned, the main hall and one of its side-wings damaged by a recent conflagration. The entire facade of the other side of the main hall (including that side's "1891"-lettering) had vanished, replaced by an immense tarpaulin, and one side-wing was now a burned-out shell. I entered the still-operational side-wing, occupied that day by a desultory trade-fair promoting new forms of public transportation to disinterested bureaucrats from Moldova and Bulgaria (the next trade-fair would be devoted to Prague's pornographic industries). Above the trade-fair stalls, the building's walls displayed cast-iron friezes envisioning 1891's industrial technologies and the concentrated corporeal figures activating them. Despite the palace's emanation of disintegrating, last-ditch survival, it still enduringly transmitted 1891's visually focused expansive technological ambition that had propelled Muybridge through Prague in his tour of that year, along with that ambition's accompanying aim to generate vast experimental auditoria and display-halls within which to engage and enthral spectators.

After leaving the palace's side-wing and circuiting its exterior, I covertly entered the deserted main hall and looked up into its vast, vertically-oriented space, surrounded at several levels by iron balconies once occupied by audiences gazing down at unprecedented technologies. One side's glass and iron facade admitted light, the other side's tarpaulin blocked all illumination. The palace's oak floor was covered in ashes and oil, upturned barrels and shattered display-cases. I slipped back out and headed down to the fairground at the Exposition's northern perimeter, boundaried by railway tracks. Everything was closed for the winter, including the site's Ferris wheel and a hexagonal, concrete-faced hall, holding a painted battle-panorama, which had opened in 1898, seven years after the Exposition had taken place. Again, I imagined a delusional phantom projection of Muybridge activating his moving-image sequences – in whatever auditorium had preceded this one, for the Exposition's duration – together with his own immersive photographic panoramas of San Francisco's urban space, prior to that city's seismic destruction of the following decade. Then I wandered down to the far edge of the derelict funfair wasteland, between the arcades, freakshows and shooting stalls.

Moving-image projection almost invariably leaves no trace, other than in the occasionally seared ocular senses or memory of its audience and the projectionist's cursory noting-down of the dates of the film-reels' delivery and dispatch. That lack of tangible trace forms an essential element of projection's status, only exacerbated in circumstances of transient, city-to-city moving-image projection. In many ways, to attempt to reconstruct or resuscitate an act of projection is itself an aberrant act, revivifying gestural movement that is now already definitively consigned to its utter vanishing, and initiating too the danger of a fall into the inescapable domain of phantom projections.

While Muybridge's Prague moving-image projection retains its distinctive spectrality and elusiveness, his principal projection event in Vienna inversely holds an explicit time and location: 5pm on Wednesday 3 June 1891, in the large Physics lecture hall of what Muybridge, in his letter to Jesse Burk and in his Chicago programme, called Vienna's "Technical High School" (*Technische Hochschule*, now the *Technische Universität*), at Karlsplatz 13, one of the city's central squares. As in Munich, he projected his work in Vienna in the context of a public event at the Technical High School and also as a private presentation for artists at the city's Society of Artists, also located on the Karlsplatz. That insistent doubling of projection venues indicates Muybridge's determination to mediate the technological and arts-focused dimensions of his work, bringing his projections to technologists, anatomists and artists. The division between the two forms of institutions and their remits was never clear-cut; in Vienna, some of the Technical High School's teaching staff were also members of the Society of Artists, and also taught at the city's arts institutions. Muybridge's audiences formed aberrant corporeal amalgams, ranging at one extreme from autocratic elderly generals who only wanted to see Muybridge's projections of running horses, to young moving-image experimenters who wanted to seize and grab what they needed of raw technological and arts innovation from Muybridge's work. His insistent doubling of venues, as in Vienna, intimates too his desire for infinite repetition, interrogating ever more deeply an identical primary obsession with the human body in movement or with that movement's disintegration, along with corporeality's transmutation in moving-image projection. Muybridge's Technical High School projection was intended as the pre-eminent public manifestation of his work in Vienna, extensively announced in the city's newspapers and more widely in those of the Austro-Hungarian Empire, on 2 June.

As with Muybridge's Berlin projection at the Urania lecture theatre, the announcements of his forthcoming Vienna projection of the following day transmit the relentless repetition of their source in a press-release which Muybridge had written himself and distributed so widely that every newspaper replicates identical anticipatory data on the event. The projection screen will be of an extremely large format. Entrance to the event will be free. Muybridge's projections of the past weeks, in "artistic" cities such as Munich and Dresden, created a sensation. Muybridge will speak in English but his commentary will be translated simultaneously into German by a fluent English-language speaker, Karl von Lutzöw, an eminent professor of art and architectural history at the Technical High School. The announcements emphasise too that the invitation for the projection had come from another renowned professor at the Technical High School, a specialist in optics and colour,

Leander Ditscheiner, who also taught at Vienna's arts academy and had recently been appointed the Technical High School's Director. In his Vienna press-release, Muybridge appears to have been intent on accentuating the lavish prestigiousness of the invitation to show his work there and on highlighting the uniqueness of his unprecedented projections, as though he had had enough in the previous months in England of projecting his work in improvised corridors or with rats scurrying on the dismal venue's floor. Although Ditscheiner's engagement with Muybridge's work evidently spanned its technological, anatomical and arts-focused dimensions, Muybridge's press-release emphasises that his "totally astounding" projections (to be shown in a Physics lecture theatre) are intended primarily for artists and "friends of art".

The eye-witness accounts of Muybridge's projections, especially those that appeared in the Vienna newspapers *Die Presse* and *Wiener Zeitung*, published on 4 and 5 June, give especially tangible evocations of how Muybridge presented and vocally commented-upon his moving-image sequences on his Central European tour. They give the event's exact duration: seventy-five minutes (though Muybridge was habitually able to project for far longer than that, and the Vienna event's duration indicates that he may have pliantly synchronised its timescale with Ditscheiner's demands). Ditscheiner himself appears to have been Muybridge's "operator" of the Zoopraxiscope on that occasion, and to have been able to project Muybridge's discs to a high standard and speed, without any malfunction, indicating his advanced technological abilities as a projectionist.

The account in *Die Presse*, which presents Muybridge as being American rather than British, highlights the corporeal aspects of his audience. Despite the auditorium's large dimensions, the spectators were heatedly crammed into the space (and as in Berlin and Munich, many spectators could not be admitted at all), and they encompassed all of Vienna's prominent artists. The projection was extremely clear, every detail of Muybridge's images visible on the large-format screen, twenty feet high and twenty feet wide. The sequences were perceived as astonishing, and as overturning the limited prior perception which spectators possessed of human and animal movement. Muybridge's Zoopraxiscope was able to run both forwards and backwards, so that particular sequences could be scrutinised in depth. The account emphasises in particular Muybridge's sequences of birds in flight – which he saw as being still in-progress and requiring future development – and their intricate correspondences with human movements. In his vocal commentary, notably in relation to the sequences of figures drawn from his Philadelphia experiments, Muybridge honed-in on muscular gestures, particularly those which indicated a loss of balance and the danger of falling. But the entranced tone of the press-account shatters at its end when it describes Muybridge's unsubtle attempts – after his projector had been extinguished, and in what the account derides as a kind of "Propaganda" – to incite his audience either to buy his *Animal Locomotion* publication or else collections of twenty-four or thirty extracted still-images from it, for art-school libraries or for artists' studios: "Every American, even the most learned, is still to some extent a businessman."[7]

The *Wiener Zeitung* account also highlights the exceptional clarity and large format of the projected sequences, emphasising (in a comparison that must have given Muybridge great pleasure) the high number of spectators and the ambitious expansiveness of Muybridge's work in distinction from that of his competitor Anschütz, whose images are described as having been shown in public "in small format and with a small number of spectators". The account asserts that: "In the artistic perspective too, Muybridge has gone much further than those who have come after him, Marey, Anschütz...". The account then reflects directly on the status of Muybridge's work as "projection", giving a detailed description of his intentional use of contrasts of light in his moving-image projections, with naked figures on horseback incorporating coloured elements moving against a black background and other, darker figures set against a "snow-white" background. As with *Die Presse*'s account, the

Wiener Zeitung emphasises the Zoopraxiscope's capacity to reverse the direction of movement, upending linear time. In his vocal commentary to his projections, Muybridge evidently gave a summary of his work's sustained focus over the course of twenty years on transmitting simultaneously the multiple dimensions of the human figure in movement. The account describes Muybridge's translator Karl von Lutzöw as standing directly alongside him, beside the projection-screen, in order to provide succinct summaries of Muybridge's commentaries; it also enumerates the names of the many artists and photographers, as well as the anatomists, technologists and "military technicians", who attended Muybridge's event.[8] In Vienna, as both accounts note, Muybridge kept his sequences of naked and near-naked figures reserved for projection at his event's end, either to avoid potentially outraging his audience at an early stage or else for a final startling revelation aimed at what those accounts regard as awe-struck spectators.

In distilling the press-accounts of his event in Vienna for the brochure of his Chicago projections, two years later, Muybridge chose particularly to highlight the Vienna projection's bursting-at-the-seams excessive corporeality noted in the account in *Die Presse*, but also used extracts from an account in Vienna's *Landwirthschaftliche Zeitung* that stressed how Muybridge's work "revolutionised" moving-image representations and constituted "a veritable treasure".[9] Muybridge's moving-image projections in Vienna, as with those in Berlin, formed a seminal art event as well as an unprecedented technological spectacle, enthralling a uniquely ecstatic audience which would directly experience his work, with its live aura of revolution, solely on that tour.

I walked through the lavish salons of what had, since Muybridge's visit in 1891, become Vienna's "Technical University" (Muybridge wryly comments in his letter of 15 July 1891 to Burk that his Vienna projection venue, along with several of his other Central European venues, had not yet "attained the dignity of name" of a university, though he was well-aware that Burk was

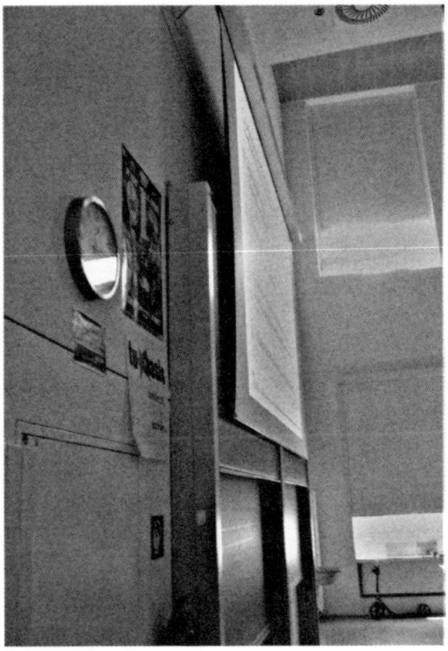

Ill. 7: Muybridge's Vienna auditorium, at the *Technische Universität*, 2020.

solely interested in whether or not the institutions in question had bought copies of the *Animal Locomotion* publication).[10] The salons' elongated windows overlooked the Karlsplatz. A reception for the inauguration of a new Director – only the fourth since Muybridge's improvised but adept projectionist, Leander Ditscheiner – was ongoing, and I was introduced by the Director to the University's architectural historian, who was in his mid-eighties and evidently knew every nail and striation of that elegant building, which had been constructed in 1857. He was able to identify immediately which auditorium Muybridge had used for his projection. It even still possessed the same identifying number, as "*Hörsaal* 8", that it had held in 1891, located on the mezzanine level between the building's first and second storeys. The historian told me that the principal alteration which the auditorium had undergone in the past 130 years was that the seating – previously in the form of individual chairs at ground level – had been replaced with a steep rake of benches. Like every auditorium in the entire building, it had been shaken by wartime bombing, but was underpinned by a resilient iron framework, as with Muybridge's Berlin venue. I asked him if the auditorium was still used, and the historian told me it was, though now for lectures in European cultural history, rather than on physics and chemistry. But it was never locked in the daytime – I could go and see it now, right now.

I walked through the building's courtyard towards the doorway leading to *Hörsaal* 8. From open windows in every direction, I could hear ongoing lectures, on the disintegration of memory, on urban wasteland contaminations, on medieval apocalyptical preoccupations. I reflected that it appeared almost too easy – after my months of intricate negotiations with the Berlin police and Prague's opacity in delivering up Muybridge's projection-venue in that city – to be simply walking into Muybridge's Vienna auditorium, which was signposted with a small digital screen. The vast ash doorway to *Hörsaal* 8 was unlocked and I entered. The high-ceilinged auditorium appeared immense at first sight. A lecture had evidently finished some time ago and several students were still in the seating rake's highest rows, finishing their notes; another lecture would start in an hour or so. I was struck that Muybridge's Vienna auditorium – unlike his now-abandoned Berlin venue – was still corporeally inhabited, 130 years after his own figure, and his projected moving-image figures, had occupied it for seventy-five minutes, rendering his audience awe-struck, before vanishing. Most of the south-west side of the auditorium, opposite its entrance, was glass-faced and looked out onto the streets behind the Technical University, the two superposed sets of vertical panes enclosed in a shuttered iron frame, currently opened so that light flooded into the space, illuminating the site of Muybridge's projection-screen. Through those panes, I could see an elegant residential building with its inscribed construction date, 1884: seven years before Muybridge had undertaken his projection here. The south wall – with its current screen equipped for digital projection, and holding approximately the same dimensions as Muybridge's own screen, installed there for his 1891 projection – was expansive, stretching up fifty feet or more to the ceiling. As his event's press-accounts indicate, Muybridge had been able to project his work, from his Zoopraxiscope located at the back of the auditorium, to a vastly detailed scale. As I stood in front of the screen, the auditorium gradually filled up behind me, and when I turned around, it was at maximum capacity, pitched corporeally in excess as it must have been for Muybridge's projection, the rows spilling bodies, latecomers standing at the sides.

I then crossed the Karlsplatz gardens – still the site of a river at the time when Muybridge had passed through Vienna on his projection-tour – towards the three wooden entrance doors to the *Künstlerhaus*, at Karlsplatz 5, built in the 1860s and located directly across that square from the Technical University. Muybridge mentions his "Society of Artists" projection – undertaken as a private event for Vienna's artists and photographers, in the days immediately following his public event – solely in its utility as having generated another *Animal Locomotion* sale. After multiple twentieth-century projects to demolish it, the building was now used as an art museum, its space mainly devoted to galleries, and the part of the building in which Muybridge had projected his work (added as a side-wing in 1882) had been assigned for the past seventy years as its cinema. I had made an appointment with the elderly projectionist, and he met me at the cinema's entrance, rather than in his projection-box, as though severely disinclined to admit visitors to that space, which still held 8mm and 16mm celluloid projectors for showing art films. When I asked him if he knew of Muybridge's event here, he at first said nothing, as though guarding secret knowledge, but I could see his hands were shaking. Finally, he whispered: "Muybridge, the killer…", and began to speak. He told me the auditorium must still hold the approximate dimensions it had in 1891, but all of its original ornamentation had been sheared-away in the course of a 1949 renovation that followed extensive wartime damage to the building. I asked him if he had already started work here at the time of that renovation, but he shook his head, as though negating time itself – no, he had arrived only in 1961. Now, in a few days, the building would close again, for at least a year, for another major renovation, so his work here was now suspended, the final projection had taken place on the previous night, and it was unknown if this auditorium would survive the renovation as an autonomous space or else become conjoined with the adjacent galleries. The cinema could even emerge intact but converted solely for digital projection.

The projectionist kept the auditorium near-darkened and I could barely make out the faded frescoes that ran along both of its sides. That auditorium, in its apparent obsolescence, transmitted nothing of the ineradicable residue of Muybridge's presence that appeared still adhered to the abandoned Urania auditorium in Berlin or to *Hörsaal* 8 across the Karlsplatz, so I asked the projectionist if he could project a film for me, and his wary demeanour abruptly transformed, his eyes now those of an obsessive, profoundly cast back once more into his mania. He nodded his head and disappeared into his projection-box. The auditorium's lights soon darkened still further – for a moment, I felt that space abruptly possessed with criminality, madness and death – and the voided screen became revivified with film, firstly in a brief sequence of numbers ("6/64"), then through images of rapidly cut actions in a cellar space, figures in livid colour moving rapidly in sexual amalgams. After the film's four-minute duration, I waited for the auditorium to be illuminated and for the projectionist to re-emerge, but he remained invisibly enclosed in his projection-box, and when I finally climbed the narrow stairway to that space and opened the door by a crack, I could see him sitting weeping under a single lightbulb, facing away from the cinema screen, his arms cradling the battered steel can of the Kren film, the Eumig-P8 projector still running.

For the month of June 1891, Muybridge travelled again from city to city in Europe in his corporeal figure's inhabitation and first embodiment of the itinerant moving-image projectionist's status. From Vienna, he headed south into Italy and projected his work in Rome, Naples, Turin and other cities. But in Italy, his projections were primarily for institutional university audi-

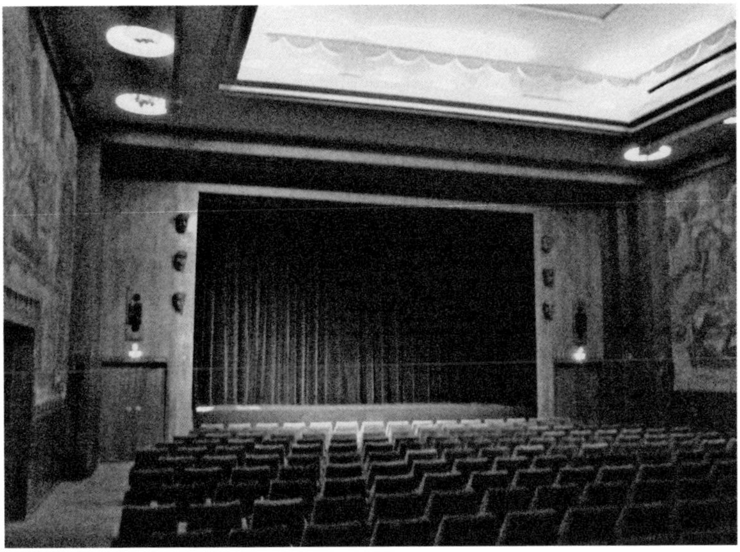

Ill. 8: Muybridge's Vienna auditorium, at the *Künstlerhaus*, 2020.

ences, and the depth of ecstatic public attention his work had generated in Germany and in the Austro-Hungarian Empire now dissipated. His overriding concern in Italy was to persuade institutions to buy copies of his *Animal Locomotion* publication. His Italian projections appear sometimes to have been improvised and certainly far less subject to meticulous planning than his earlier Central European projections (with their intensively controlled media regime); they left no press-accounts. Muybridge's own account in his 15 July letter to Burk of his Turin projection, articulated as an incident representing yet more triumph for Muybridge, also contrarily lets slip the fragility of his improvisational projection-travel, and his status in Italy as a transient moving-image projectionist arriving in a particular city without any certainty that he would be projecting his work there at all: "At Turin I presented my letter of introduction to Professor Mosso at 9 in the morning, and was asked by him to return to his lecture room at 3 at which time I found assembled there a large number of distinguished Turinians...".[11] Muybridge asserts that his vocal commentary to his Turin projection event had to be translated by Professor Mosso who "very kindly rendered my remarks to the audience in his own admirable style, as I was not quite familiar with the Turin dialect", indicating that for his other projections in Italy, Muybridge had spoken in Italian, which appears to have been the only language other than English that he really knew, despite his claims to Burk of an omniscient all-language fluency.

Muybridge discloses to Burk that his itinerary in Italy had been subject to short-term blackouts or suspensions of projection that appear antithetical to his habitual relentless determination to project, and which suggest that his tour's last projections in Italy possessed an accumulating aura of disintegration, in which projection's obsessionality finally stalls and fragments, beyond its propulsive spatial expansion through Muybridge's tour across Europe's cities that had incited maximal awe; moving-image projection, after its illuminated intensity, is invariably subject to a terminal deliquescence. Muybridge reveals to Burk with evident hesitation that he took a holiday after his projection in June 1891 in Naples (the city in which Muybridge's one-time ally and competitor Etienne-Jules Marey habitually spent his summers, though Muybridge omits any mention of him in his letter to Burk). Even Muybridge's account of his holiday is presented as a spectacle of endurance and ocular observation: "I decided at Naples that as I had not devoted much time to pleasure since I left the US, to have two days of holiday, so duly ascended to the summit of Vesuvius and examined the workings of the craterial interior; and also wandered through the once animated streets of Pompeii."[12] Muybridge's profound attachment to his Zoopraxiscope projector must have made Burk wonder if he had carried it to the summit of Vesuvius with him, rather than leaving it behind in his cheap-hotel room. Muybridge's account of his wanderings through Pompeii anticipates his oncoming catastrophe at the Chicago Exposition two years later, when a panorama spectacle depicting the destruction of Pompeii would abruptly supplant his planned programme of projections, mid-Exposition, taking over his

Zoopraxographical Hall and ejecting Muybridge from his projectionist's status and his own moving-image auditorium, as though as a belated retribution for his decision to take a holiday in Naples from projection's exacting duties.

From Italy, Muybridge returned to his tour's base in Berlin, from where he wrote his account of it to Burk. His Central European moving-image projection tour of the previous four months had evidently electrified Muybridge, especially through the rapturous responses to his Berlin, Munich and Vienna projections, and now led him to begin to envision a far more expansive tour, to India, Australia and Japan (in 1891, India and Australia were both still part of the British Empire). In that planned tour with its vastly ambitious urban scope, he would become a global moving-image projectionist, transient across immense distances, traversing countries with their own unique performance lineages of sequential corporeal movement, and momentarily inhabiting cities in which his projections would form miraculous, beguiling events. Those tentative tour-plans were developed for the following year, until October 1892, when Muybridge abandoned them in his overriding desire to create his own moving-image auditorium for projections to the potentially immense audiences of the Chicago Exposition.

At the end of his letter to Burk, Muybridge's grandiose enumerations of the many Central European cities in which he had projected his work – with extreme success suddenly lapse, as though subject to an abrupt technical malfunction. He appears contrarily now to want to render that projection-tour hidden or lost from knowledge and memory – thereby suspending and accentuating it in its momentary and precarious glory, and in its immediate ocular and corporeal live dynamics – asserting that he "would prefer nothing being said about it in the [US] [news]papers at present when it will do no good".[13] Instead, he turns to the mundane business of the sales of *Animal Locomotion* – the "principal object of this letter" – instructing Burk to facilitate the sending of promotional material about the publication to potential purchasers. In the letter's last words, he then dismisses Burk with caustic disdain, as though conscious of having revealed too much – of himself, and of his Central European projection-tour – through his own words: "Hoping you are well and will enjoy your vacation to return with renewed vigor to your post."[14] Muybridge would leave Berlin later that summer, travelling first to England, then sailing on 25 November from Liverpool to New York.

Notes

1. This chapter is mainly drawn from archival research into materials at the Muybridge Collection at the Kingston Museum, especially Muybridge's own personal "scrapbook" and other documents relating to his Central European projection tour, as well as from visits to the surviving auditoria of Muybridge's 1891 projections. I'm grateful to everyone, particularly Stefan Donath, who helped me in mapping Muybridge's tour and in enabling me to enter those surviving auditoria in which he projected his work in 1891, as well as to the staff at the many libraries and archives which hold press-accounts of the responses to Muybridge's projections in European cities.

2. Ernest Webster, manuscript letter, 18 November 1931, Muybridge Collection, Kingston Museum, Kingston-on-Thames.

3. Eadweard Muybridge, manuscript letter, 15 July 1891, Muybridge Collection, University Archives and Records Center, University of Pennsylvania, Philadelphia.

4. Ibid.

5. Muybridge, *Descriptive Zoopraxography* (programme for his Zoopraxographical Hall projections), University of Pennsylvania, Philadelphia, 1893, page 33.

6. Muybridge, *Descriptive Zoopraxography*, pages 32-33, and Muybridge, personal scrapbook (1894-1904), unpublished, conserved at the Kingston Museum, Kingston-on-Thames, page 167. In *Descriptive Zoopraxography*, Muybridge uses the translation "last corner", and in his scrapbook, "extreme corner", the latter of which is a better rendering of the German-language phrase in the original review of 17 April 1891 in the *Münchner Neueste Nachrichten*. Other newspapers gave accounts of Muybridge's Odeon projection-lecture of 16 April, such as the *Allgemeine Zeitung*, which also reported on Muybridge's private event in Munich on 10 April for artists affiliated with the city's Society of Artists.

7. Anonymous account in *Die Presse*, Vienna, 4 June 1891, page 9.

8. Anonymous account in the *Wiener Zeitung*, Vienna, 5 June 1891, page 7.

9. *Descriptive Zoopraxography*, page 30.

10. Muybridge, letter of 15 July 1891.

11. Ibid.

12. Ibid. Muybridge had left the USA over two years earlier, on 26 January 1889, sailing from New York to Liverpool.

13. Ibid.

14. Ibid.

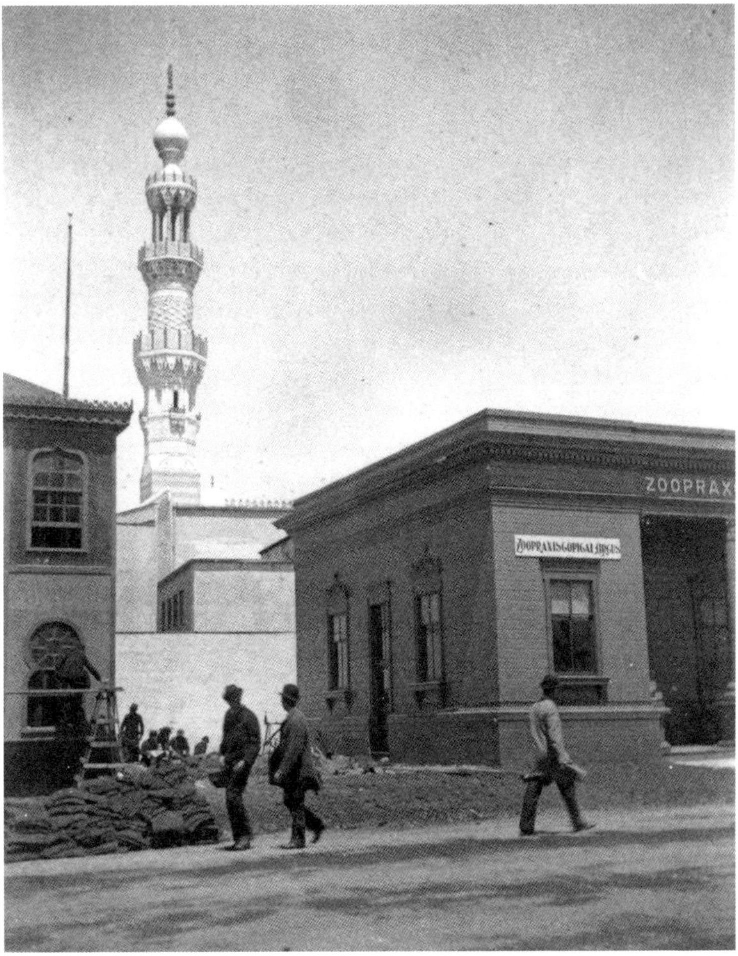

Ill. 9: Untitled photograph of Muybridge's "Zoopraxographical Hall" under construction at the Chicago World's Columbian Exposition site, 1892-93

Muybridge's Chicago Exposition Projections, 1893

The Chicago World's Columbian Exposition of May to October 1893 formed an extravagant urban dream of the USA's future, and of global futures, cast in architecture, spectacle, corporeality, knowledge and entertainment, and located as an immense sub-city at Chicago's southern periphery, between its newly ascendant mansion suburbs such as Washington Park and the expanse of Lake Michigan. Alongside millions of visitors from other countries, a large percentage of the young population of the USA would be dispatched to the Exposition Park – just as the young populations of future locations of Expositions, such as Japan's Osaka Exposition in 1970, would also dutifully attend them – to spend a week or so assimilating new and even unprecedented developments in art, technology and science, viewing exhibitions and attending lectures in the specially constructed pavilions installed around a network of lagoons. The spacious exhibition pavilions were often named "Palaces" – as with the Palace of Fine Arts, the only still-surviving Exposition Park building – to imbue them with the distinctive prestige similarly accorded, two decades on, to the vast and ornate "film-palace" cinemas of the USA's major cities. But directly alongside the Exposition Park, an askew annex to it was situated, devoted to forty-one "concession" sites hired by entrepreneurs, and that strange and unruly zone formed, in many ways, the urban antithesis to the main site's more instructive space, with its grand pavilions' educational programmes including revered speakers such as Swami Vivekananda and other religious philosophers. Along the Midway Plaisance annex, by contrast, fee-paying visitors could watch the sexual spectacle of the Tunisia pavilion's "dancing girls", walk along an outrageous "street in Cairo", ascend a vertiginous fairground Ferris wheel, and be sensorially entertained by new technologies of vision and sound. The Midway Plaisance was the site where Muybridge, after suspending his moving-image projection tours of the previous decade, commissioned the construction of the first auditorium exclusively designed for the projection of moving-image sequences, his "Zoopraxographical Hall". At the end of the Exposition, all that would remain of its vast expanse of educational palaces was a burnt-out wasteland, and Muybridge's own moving-image auditorium – no longer his, after the disaster that assailed him in Chicago – would be destroyed without trace too.

From the first planning phase of the Exposition following the initial struggles of Chicago's urban figures of power to locate it alongside their own city, rather than its fierce competitor, New York, there had been the determination that the event would encompass laboratories for experimentation in human vision. Once the Exposition had been awarded to Chicago, its planners immediately decided on the division of the site into the principal area of its educational palaces, and an additional, finance-generating area (extruding horizontally from its western edge) for sensorial experiments and entertainments, semi-autonomous and only nominally under the Exposition's authority. The Exposition's *Report of the Committee on Grounds and Buildings*, from 31 August 1891 (almost two years before the event's opening), indicated that the Midway Plaisance could, alongside the display of outlandish human bodies, incorporate venues for projections and panoramas: "Many applications have been made for locations for buildings by private parties, companies and syndicates. As a general policy it has been deemed wise to place these exhibits showing the manners and customs of peoples in Midway Plaisance. Among the displays already proposed are a Bazaar of All Nations, a Street in Cairo, an Indian Village, Turkish Village and other similar representations of life and manners in foreign countries... There will be located in Midway Plaisance panoramas, cycloramas, etc."[1]

Muybridge was a relative latecomer to that application process, since it was only in October 1892, seven months before the Exposition's opening, that he applied for his own Midway Plaisance concession, positioned at the central point of the Midway avenue's northern side, to be occupied by a building of sixty by fifty feet in size; he asserted that the purpose of his concession would be to "give lectures" on his experiments in recording animal locomotion, in addition to selling related merchandise, and he paid a deposit of $1,000 to secure his site, along with a further $6,000 in bonds in the form of guarantees, and agreed that one third of all of his takings would be handed over to the Exposition authorities.[2] Muybridge preferred to give the impression that he had been directly invited by the Exposition authorities to participate in the event, rather than having himself to pay for a concession in the Exposition's detrital hinterland as an entrepreneurial "private party" and to meet all of the costs for his auditorium's construction. He highlights that emphasis on receiving an "invitation" (rather than on seeing press announcements of the opportunity, as appears to have been more likely) in the narrative he wrote for the programme sold at his Zoopraxographical Hall during the Exposition: "In the Summer of 1892 while the Author was in California, preparing for a Lecturing tour through Australia and India, he received an invitation from the Fine Arts Commission of the World's Columbian Exposition to give a series of lectures on ZOOPRAXOGRAPHY in association with the Exposition now being held in Chicago."[3] He travelled to Chicago in October 1892 to sign his Exposition contract and to commission a long-established Chicago-based architect, Thomas W. Wing, to construct his projection auditorium to his own specifications.

Following his Central European moving-image projection tour of March to July 1891 – and his exposure in Europe's cities to ecstatic, awe-struck audiences – Muybridge's project ambitiously to expand that experiment in moving-image itinerance, by undertaking a more global and vast projection tour, occupied him for most of the following year. Alongside its British colonial dimensions in India and Australia, where Muybridge would be able to perform his vocal moving-image commentary directly in English, Japan appears always to have been intended as the culmination of his planned tour; Muybridge often evoked his admiration for Japanese art in his projection commentaries, and Japan's then-emergent engagement with photographic experimentation and its new eager openness to technological innovations from Europe and the USA must also have been attractive for Muybridge, who above all desired the adulation of spectators for whom his work would appear entirely unknown and unprecedented. He described his planned moving-image projection tour of India, Australia and Japan as his "far Occidental expedition", thereby evoking its exploratory and exacting dimensions, as though it involved the expeditionary scaling of K2, Mount Kosciuszko or Mount Fuji, in the way that he had climbed Mount Vesuvius during his 1891 European projection tour.[4] No pre-existent infrastructure for his planned global tour existed beyond the scattered presence of entrepreneurs, notably in the principal cities of "British India", who arranged visits of theatrical spectacles from Europe. Muybridge's tour would demand the prior identification of auditoria suitable for moving-image projection in each city, as well as the dependability of travelling "operators" able to transport and activate his delicate Zoopraxiscope, and the hiring of translators in Japan.

His planned tour appears an infinitely excessive operation, corporeally arduous in the extreme, and intended to originate inexistent audiences who would need to be conjured-up by the projections themselves, but Muybridge evidently planned to accomplish that tour. Between his return from Berlin to the USA in November 1891 and his commitment in October of the following year to devote most of 1893 to his Chicago Exposition projections, Muybridge only intermittently projected his work for public audiences, in the form of one-off events in the first months of 1892 in New York and San Francisco, and otherwise was engaged in those tour plans, intended to propel his status as an itinerant moving-image projectionist to maximal dimensions. His other preoccupation of that interval time was to experiment with photographing sequences of insects and birds in flight.

Muybridge finally decided to prioritise his Chicago opportunity, and thereby abandon or defer his global tour plans (in his Chicago programme he writes that he has only "delayed" them, though solely the ambition to project his work in Japan survived his Chicago calamities), at a point when those plans had clearly reached an advanced stage. His renunciation of his own seminal itinerance in moving-image projection forms a momentous one, paralleling the future decisions of worldwide moving-image projectionists across the subsequent decades who chose to relinquish their itinerant city-to-city transits and instead become the employees of cinemas, confined spatially to projection boxes. Alongside his capacity to envisage the creating in Chicago of a moving-image auditorium for his work, Muybridge must also have been enticed by the delirious ways in which the Exposition's pre-publicity formulated its dreams of the future USA that was to be instigated by the event: visions of new visual technologies and imageries, of transformed corporealities and urban environments, and of spectacular architectures for those dreams' realizing. While Muybridge had received almost total adulation for his work on his 1891 European tour, that concentrated reception of awe had always eluded him in the USA, where his projections had been intermittent; the Chicago Exposition, with its durational expanse that gave him the opportunity to undertake multiple projections on one site (he initially planned at least 300 projections over six months), within a highly mediatised technological environment that could transmit knowledge of his unprecedented work worldwide, evidently convinced Muybridge to prioritise fixed space over volatile itinerance.

His Chicago opportunity may also have appeared to provide Muybridge with the awareness of a uniquely last-ditch, last-chance moment finally to generate the recognition he desired in the USA, before his work's imminent engulfing and casting into oblivion by his younger competitors' own innumerable moving-image innovations. He installed his work in Chicago, in a way that prefigures moving-image or performance installation in contemporary art contexts. The element of obsessional repetition that had always been pivotal to Muybridge's multiple-camera experiments in human and animal locomotion also proved central to his Chicago projection plans; he intended always to project the same moving-image sequences and to present the same invariable vocal commentary, many times each day.

Muybridge's dilemma of October 1892 – whether to expand his itinerant projections globally and indefinitely, or else to install his moving-image sequences and vocal commentaries within an auditorium of his own architectural design, in an experimental city, for a fixed period – also determined the future of moving-image culture, of film, and of digital projections both within installation environments and as the generating element of all cities' image-focused external infrastructure, through immense projecting screens attached to buildings' facades. Muybridge appears to have decided that his corporeal and urban obsessions required spatial stasis, and his contract with the Exposition authorities signed-away his moving-image projection itinerance, as least for the event's duration (and in the end, that curtailed agreement would go awry). Once his contract was signed, Muybridge was compelled to envision what the optimum spatial environment would be for his moving-image sequences' projections, both in its exterior and internal aspects. He had to create the first-ever auditorium for the projection of moving images to public audiences.

Muybridge had previously generated the audience for his work through his intensive regime of press-releases and via press accounts, by the conjunction of his own preoccupations with those of his spectators in their mania to witness corporealities in movement, and also through the aura of unprecedented innovation that had attached itself to his work from the early 1880s and which had been at its zenith during his tour of Central Europe's cities. At the Chicago Exposition, with his auditorium positioned against the Midway Plaisance's many other spectacles, Muybridge faced direct competition for the sensorial capacities of his prospective audiences in the adjacent pavilions' unknown ocular and olfactory attractions, with their sexualised

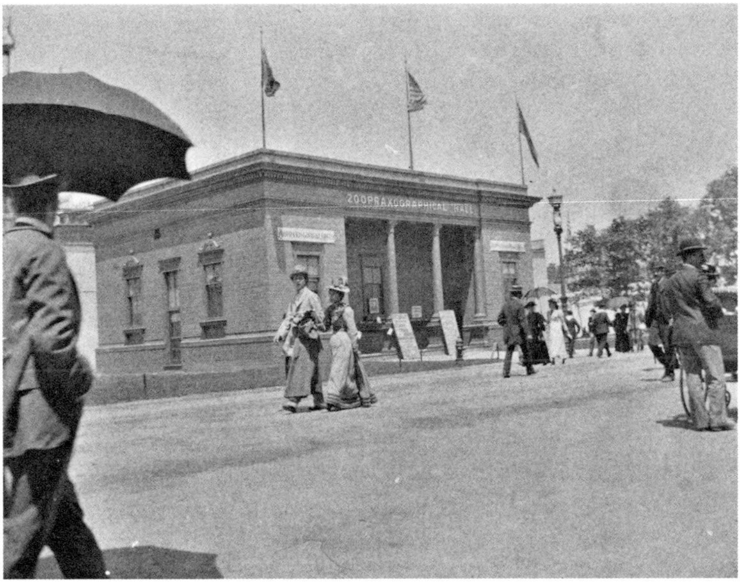

Ill. 10: Untitled photograph of figures walking on the Midway Plaisance avenue at the Chicago World's Columbian Exposition, in front of Muybridge's "Zoopraxographical Hall", 1893.

dances, and in the chaotic uproar of that zone existing beyond the Exposition's instructional main area. Muybridge's auditorium formed a contrary anomaly within the Midway Plaisance's overspilling chaos – with alleys simulating exotic bazaars, and panoramas evoking calamity – but also functioned spatially as the crucial axis from which the entirety of that excessive life and movement was mediated. Although Muybridge's professional competitors such as Edison and Anschütz attended or participated in the Exposition, they could not compete with the directness and ambition with which Muybridge conceived of a moving-image auditorium to project his own work.

Muybridge's Zoopraxographical Hall worked on the principle that, once an auditorium had been created for the projection of moving-image sequences, an audience would irresistibly want to assemble outside, several times a day, pay a fee and be admitted inside to re-assemble there, body in intimacy with body, collectively facing a large screen attached to the wall opposite the entrance, and to remain in a static position with their eyes focused on the screen, seated or standing, in order to experience the projection of moving images. Unlike the venues in which Muybridge had projected his work in recent years, such as those in Berlin and Vienna, that building had been created for no other purpose whatsoever, and would lapse without its moving-image audience. In Muybridge's auditorium, the projections' creator was also the striking corporeal figure who now stood alongside the screen, vocally commentating on his sequences. That act of exhibition, undertaken by Muybridge from the first days of May 1893, held its unique distinction of unleashing the compulsive corporeal and sensory experience of public moving-image projection, that would be exercised for a century and more, in future-oriented volatility.

That future history of moving-image auditoria – in their multiple manifestations, and each with their own particular projection-box environments – extends from Muybridge's Zoopraxographical Hall through architectural experimentations in cinematic urban space such as those of Downtown Los Angeles' 1910s-20s avenue of luxurious, technologically advanced cinemas, "cathedrals" and "palaces" of film, many of them designed by the renowned architect (solely, an architect of cinemas) S. Charles Lee, via detrital city-edge clapboard or corrugated-iron spaces of improvised projection of pornography or exploitation cinema, and on to contemporary digital immersive auditoria, in a non-linear and vitally malfunctional historical sequence in which the "disappearance" and "vanishing" of cinemas, and of their projectionists, is always an imminent event. All of those projection auditoria originate from the moment when the first spectator paid a small fee, twenty-five cents, and entered Muybridge's Zoopraxographical Hall on the Midway Plaisance.

Muybridge's previous audiences had been extracted or adapted from those of pre-existent areas of attention, especially from engagements with art and technology. For the subsequent fourteen years or so (until film's ascendant industries finally caught up with Muybridge's innovations, with the urban mass-construction of cinemas), the audiences for moving images were often also extended from existing variety-spectacle audiences. But to materialise spectatorship on the Midway Plaisance, Muybridge had to generate an immediate audience without precedent. Streams of avid prospective spectators would constantly be walking along the Midway Plaisance, deciding which of the forty-one concessions they would enter, but as a potential audience for Muybridge's Zoopraxographical Hall, they would need to be adroitly directed inside it. He formulated the construction of large inscribed placards to be positioned at his venue's entrance, and prepared posters to be attached to its facade, announcing the times of projections and their cost. He commissioned a large-scale photographic portrait of himself to be affixed to the outside of the auditorium's main entrance door, and though his audience would almost invariably never have previously heard of him and would not recognise that portrait, they would assume that the attraction to be experienced there had an instigator: a mysterious figure with a long white beard, staring out at them. The grandiose name "Zoopraxographical Hall" was inscribed in large letters above the building's entrance with its four Roman columns, but Muybridge also deployed an alternative, audience-directed name for his auditorium that undercut the other's gravity: "Zoopraxographical Circus", announced on banners above the ticket-kiosks at either side of the building's facade. With its evocation of ritualised animality, that doubled name also conjured the strange and omniscient "ringmaster" of projection portrayed at the building's entrance, Muybridge, along with an inflection of Roman-arena enticements of aberration and violence.

Muybridge's fixed-space conception of his Zoopraxographical Hall embodies unique and vital innovation, but simultaneously also transmits Muybridge's exhaustion after his decade of gruelling work as an itinerant moving-image projectionist. The Exposition authorities constructed a large boarding-house for Midway Plaisance exhibitors, located directly alongside Lake

Michigan, and Muybridge's anticipated travels of the Exposition's six-month duration would be reduced down to the thirty-minute walk from that boarding-house, around the Exposition's official palaces, under the railway tracks that brought visitors to the event, through the turnstiles of the Midway Plaisance and halfway across the avenue (a small "sliding" railway with regularly sited stations, initially used to assist exhibitors in installing their attractions, was available for anyone who was too incapacitated or weary to traverse the Midway Plaisance on foot, and sedan chairs could also be hired for 75 cents) to his Zoopraxographical Hall. That short boarding-house to auditorium journey comprised the extreme miniaturisation of Muybridge's envisaged global "expedition"-tour of projection, and thereby may have formed a seductive impetus for his last-ditch project at the Chicago Expedition. Once he had signed his contract, Muybridge had to stay put, to renounce itinerance, and the format of the Exposition constituted a rigid template for that freezing into installation of his previously mobile work. Muybridge's subsequent reticence on his Chicago Exposition experiences, across the final decade of his life, exuded an exhausted silence. He would subsequently avoid and elude those experiences in Chicago, displacing his point of departure when he subsequently returned to Kingston-on-Thames, and solely evoking the Exposition's own official prestige when he referred to it in the context of his final UK projection tour. His time in Chicago barely figures in the great scrapbook of achievements which he spent most of his final decade compiling; some of the documentation he had amassed of his travels of 1891-93 was accidentally lost during the era of his return to Kingston-on-Thames in 1894, perhaps by his Philadelphia associate, Jesse Burk, but Muybridge also appears to have directed a final gesture of oblivion towards his work in Chicago.

I travelled to Chicago and visited the exact location of Muybridge's Zoopraxographical Hall, within a sunken grassgrown sports-terrain adjacent to the Woodlawn Avenue viaduct whose predecessor appears in the original mappings and photographings of the 1893 Midway Plaisance attractions. The icy open-air projection ground had no trace or marking whatsoever of its moment of holding the originating moving-image auditorium, which formed the first "vanished" cinema, razed more comprehensively even than the earthquake-decimated, haunting city-ruins, palaces and cathedrals of Antigua Guatemala which Muybridge photographed after fleeing from California to Central America in 1875 following the scandal of his murder trial for shooting-dead his wife's lover. From the Zoopraxographical Hall's location – where Muybridge had planned to project his work in inexorable repetition – I walked eastwards, under the railway tracks into Jackson Park and past the splendour of the Exposition's surviving Palace of Fine Arts, within the event's main site, where Muybridge may well have believed that his projections rightfully belonged, then over to the lakeside location of the exhibitors' boarding-house, and to the Lake Shore Drive parapet extending over frozen Lake Michigan.

For his Zoopraxographical Hall's construction, Muybridge hired a pliable architect without a distinctive style of his own, in a parallel way to that in which musicians hire studio producers or engineers with blank, undistinguished reputations, so that they can impose their own plans or aberrations totally. Thomas W. Wing was almost sixty years old at the time when Muybridge hired him and had been based in Chicago for over twenty years. He was known primarily for a mansion he had designed at the start of his career in the city, the Raber House, built in 1870 for a politician and entrepreneur of German origin, John Raber, forming the centrepiece of what had then been a splendid estate of parkland, alongside other estates owned by newly-wealthy entrepreneurs in the Washington Park district of southern Chicago, not far from what would become the Exposition's Midway Plaisance site. The district's prestige then disintegrated, and Wing returned to the Raber House in 1892, shortly before the Exposition, to convert its four storeys into small apartments, shearing off the house's previous ornamentation and side-wings so that it appeared strangely elongated and awry. During that era, he was still building grand houses, but now in the ascendant northern Lincoln Park district, and he completed work on a house in that district's salubrious Deming Place, shortly after he had constructed Muybridge's Zoopraxographical Hall.

Wing had worked for several years for the architectural company Burnham & Root, which won the contract to become the Exposition's primary architects; one of the two senior partners, John Wellborn Root, died in 1891 at the very moment when the Exposition plans were taking shape, and Daniel Burnham then took the project forward with other architects, implementing the main site's neoclassical palaces and boulevards (Burnham later specialised in skyscrapers, and designed New York's Flatiron Building). The designing of buildings for the Midway Plaisance annex-site was no concern of Burnham's prestigious company, and was left to far lesser hands. Wing was working independently again by the time he was hired by Muybridge, though his previous association with Burnham & Root probably led to his receiving the commission. All of the Exposition's buildings were constructed only to survive for its six-month span, and were expected to be demolished or intentionally destroyed by fire soon afterwards. As a result, the Zoopraxographical Hall was a faster, shoddier project than Wing's more habitual work around 1892-93 of constructing large houses. Even so, it was only through his work on the Zoopraxographical Hall that Wing was remembered on his death in 1916. His obituary in the *Chicago Tribune* (mis-spelling Muybridge's invented first name) solely mentions that building in its account of Wing's undistinguished career, and also indicates how the Zoopraxographical Hall's reputation as the initiating moving-image auditorium was perceived, twenty years or so after its curtailed existence: "Thomas W. Wing, architect, died on Wednesday at 800 George Street... In 1893 Mr. Wing was the architect of what was probably the first exclusive motion picture theater in the world – the frame structure on the Midway Plaisance for the display of Eduard Muybridge's 'animated pictures'."[5]

Once Muybridge had commissioned Wing to build the Zoopraxographical Hall during his visit to Chicago in October 1892, for completion by the Exposition's opening on 1 May 1893, Wing gave him a preliminary architectural drawing, which Muybridge then reproduced in his *Descriptive Zoopraxography* programme for the Exposition. Muybridge's conception of his moving-image auditorium as a form of extravagant Roman temple was rendered by Wing in his drawing through a spacious and autonomous building, exaggerated in size from its actual dimensions, with four classical pillars, two at the front facade's far sides and two intermediate pillars framing the venue's entrance door. Wing drew lush, imaginary foliage around the building which would be entirely absent from the actual Zoopraxographical Hall, built up-tight in immediate proximity to the jolting presence of the "Street in Cairo" and other adjacent Midway Plaisance attractions; he added a dome above that foliage, probably evoking the Exposition's Administration Building, and suggesting that the main site's palatial buildings were far closer to Muybridge's auditorium than they actually would be. Wing also drew many human figures around the well-attended building, both entering and exiting it in eager streams of imagined moving-image spectators, and also loitering around it in groups as though unsure whether to enter. Wing also appears to have copied ornamental elements of the Zoopraxographical Hall's design from a house constructed in the Lincoln Park district in 1892, the Jacob Gross House, designed by the more celebrated architect Edmund R. Krause.

In many ways, Wing served as Muybridge's architectural "operator" for the Zoopraxographical Hall's spatial environment, in a parallel way to that in which innumerable anonymous projectionists – and one known one, Ernest Webster – held the status as his operators of the Zoopraxiscope, substituting for Muybridge as he stood alongside his screen for his vocal commentary on his sequences. Muybridge's idiosyncratic conception of his projection auditorium's exterior evidently had to be followed without deviation by Wing, just as Muybridge's surrogate projectionists unerringly complied with the speed and sequencing set by Muybridge, articulated in his commentary's vocal emphases, during his moving-image tours. Muybridge possessed an acute sense of the environment in which he wanted his work to be experienced, even in its projection tours' itinerance, and the Zoopraxographical Hall presented his sole opportunity to be exclusively in control of that environment, rather than adapting pre-existent auditoria such as that of Berlin's Urania institution or the Vienna Technical High School's lecture-hall. Wing executed his commission in low-quality splendour, and duly on-time for the Exposition's opening and for Muybridge's first fixed-location projections of his work; Muybridge wrote in his programme that he had returned to Chicago "to find a commodious theatre erected for this special purpose on the grounds of the Exposition", as though that auditorium had somehow materialised by itself without Wing's work.[6]

Muybridge's vanished Zoopraxographical Hall left no trace on the Midway Plaisance, but two of Wing's buildings still survive in contemporary Chicago – the abandoned Raber House, located in a lethal depopulated urban wasteland, and the still-inhabited Frederick J. Lange House, which Wing built in 1893, very shortly after completing the Zoopraxographical Hall – and both, in their contrary forms, conjure elements of Muybridge's first projection auditorium. The Raber House, converted during the era of the Exposition from its original mansion dimensions to streamlined apartments, indicates Wing's adroit capacity to undertake efficient work on an extravagantly conceived building, as he did for the Zoopraxographical Hall; his Lange House conserves and adapts many of the architectural elements of the Zoopraxographical Hall, including the four Roman columns at its entrance, as though – devoid of ideas himself – he were transferring Muybridge's distinctive conception of his projection auditorium to the domain of human inhabitation, as a form of prototype live-in cinema, of the kind occupied by obsessive projectionists of later decades who could not bear to be separated from their projection-box environments.

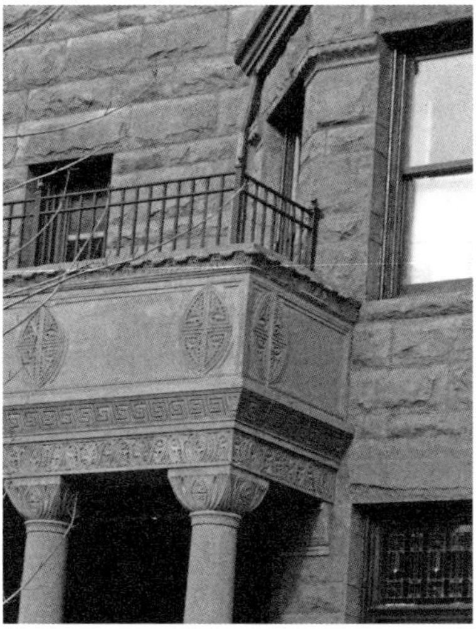

Ill. 11: The Lange House of Muybridge's architect Thomas W. Wing (1893), 2020.

I travelled to the Lincoln Park district of Chicago to visit the Lange House, in a street of large, gardened residential buildings, most of them detached. As soon as I saw it, the acute similarity of Wing's preliminary drawing – rendering Muybridge's conception of the Zoopraxographical Hall – with the Lange house's facade was startling. The house's ochre brickwork and ornamentation above its pillared entrance had been maintained in perfect condition, through the combination of its subsequent care and the relative luxuriousness of Wing's commission. The large dimensions of the bricks mirrored those of Muybridge's building. I had made an appointment with the owner to look inside, and the ground-storey interior extended for vast high-ceilinged room after room in the direction of the garden, in what appeared an acute mismatch with the house's narrow front facade; even so, the absence of any surviving evidence of the Zoopraxographical Hall's interior made the Lange house's own interior appear strangely dislocated, and I headed back out again to the facade that seemed to propel that house into a time-annulling conjunction with Muybridge's auditorium, as it appears in Wing's drawing of it and in its Exposition-era photographing.

That night, I decided I wanted also to see Wing's Raber House, though I had been warned not to approach it, either in daylight or in darkness. While the Lange house's surviving form embodied the Zoopraxographical Hall in its near-simultaneous adaptation of Muybridge's envisioning of his projection auditorium, the Raber House, by contrast, formed the time-reversed haunted-house detritus of Muybridge's conception of moving-image space, transmitted through Wing as its spectral medium. The Raber House was the last-surviving of the many 1870s mansions of Washington Park commissioned by its new population of German origin, and appeared not to have been razed in recent decades more through extreme negligence rather than any active desire for its preservation. After Wing's streamlining and conversion of the building into apartments – during the era when its inhabitants could have walked, in thirty minutes or so, via sedate avenues, to the Midway Plaisance's western periphery (where a secondary set of turnstiles enabled entrance to visitors who avoided the Exposition's main site altogether), and then on, to experience the moving-image projections of the Zoopraxographical Hall – the surrounding district's population abruptly changed several times, and the space around the Raber House was bought-up, so that its landscaped gardens disappeared. The district reached its maximal population in the 1950s, then spiralled down into acute poverty, violence and dilapidation, becoming a site of relentless gang-warfare. Heavily-armed street-gangs such as the "Devil's Disciples" took control of Washington Park, posting snipers on rooftops to facilitate industrial-scale crack-cocaine operations. Decades of drug-warfare finally exhausted the district's population, which moved out, leaving gutted houses that were then demolished, and creating a great wasteland of black earth and debris around the Raber House; all that now remained there was a minuscule population that refused to leave, alongside last-ditch drug-dealers. The Raber House itself was long-abandoned. Even so, that building had to be ocularly sieved for its architectural traces of Muybridge's Zoopraxographical Hall.

When I asked the Chicago-based theological scholar Jeremy Biles to drive us for a nocturnal excursion to the Raber House in his ramshackle car, he looked aghast, cursed, and told me "that's the number one most dangerous site in Chicago"; it was the dead of winter in the city, deep snowfall turning to icy rain and high winds. We drove southwards along Lake Shore Drive to the Exposition's Palace of Fine Art, then turned westwards from Hyde Park, passing the site of the Midway Plaisance. Eventually, the houses disappeared, and all that was left was a razed zone. Jeremy had to slow down in the pelting iced rain as we turned onto South Lafayette Avenue, and he wryly noted that the dead-of-night extreme weather conditions might discourage the prevalence of car-jackers, at least for a few minutes. I peered through the windscreen to search for the house but it soon appeared of its own accord: the Raber House was now virtually the only surviving building in the darkened narrow avenue, located on its right side, and we zigzagged to a tentative skidded halt on the tarmac. The house extended vertically above us, appearing immensely tall and askew, as a malicious apparition, set back from the pavement, a few straggling trees on either side, its once-majestic marble entry-stairway gone, every window and its strangely elongated door metal-shuttered. We opened the car's doors and stepped out, leaving those doors wide-open for a quick getaway. The house was engulfed in silence, but somewhere in the wasteland behind it, two projecting beams from unknown sources extended directly upwards into the rain-slanted sky. I took a few spectral photographs with Jeremy's malfunctioning smartphone, having forgotten a camera in the furore of that last-moment decision to visit the Raber House. In its wrecked, tangential evocation of Muybridge's Zoopraxographical Hall – via Wing's simultaneously excoriated and gutted architecture – it exuded all of the amassed future devilishnesses, disintegrations and erasures of moving-image locations and of film's haunting projections. We could hear a few

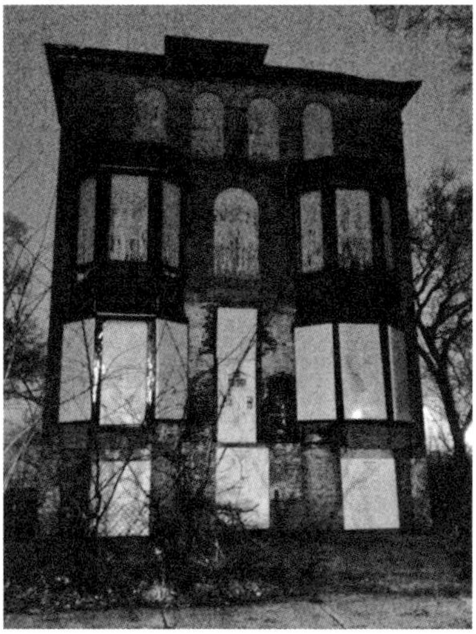

Ill. 12: The Raber House of Muybridge's architect Thomas W. Wing (1870/renovated c.1892, now derelict), 2020.

stealthy shouts from behind the house, and Jeremy whispered that it was definitely time to leave, right now. We got back into the car, he accelerated away, and a few moments later we were out of South Lafayette Avenue, heading past King Drive station on the way back to the Midway Plaisance.

The exterior of Muybridge's Zoopraxographical Hall on the Midway Plaisance was photographed on numerous occasions over the course of the Exposition, but always at an awry tangent, or as a fragment, or accidentally, or in the background of other photographs, or as an incidental detail in a panorama of the Midway Plaisance's attractions; no known photographs of its interior exist. In the immense collection of photographs of the Midway Plaisance in the archives of the Chicago History Museum, the facade of the Zoopraxographical Hall appears around fifty times. No photographer appears ever to have intentionally documented the building, through a frontal image, and the only representation of the auditorium that focuses entirely upon it is Wing's idealised drawing from October 1892, before it had been built, and in which it appears in splendour with no other attractions' pavilions alongside it. By contrast, many of the photographs showing the Zoopraxographical Hall reinforce a sense of its rawly intimate proximity to the far larger Street in Cairo attraction directly behind it and the "Persian Concession" a few feet alongside it on its west side. It directly faced the much larger "Moorish Palace" on the other side of the Midway Plaisance's wide pedestrian walkway. The Zoopraxographical Hall's adjacent buildings, predominantly featuring extravagant domes and ornamented minarets, made it appear negligible, its relatively staid construction and Roman columns outlandishly misplaced in its Midway Plaisance location that emphasised a concocted visual and olfactory culture of Middle Eastern and North African sites. On its eastern side, the Zoopraxographical Hall was bordered by a small empty plot (evidently one for which no exhibitor had been found) and then by the Woodlawn Avenue viaduct carrying pedestrians and vehicles over the Midway Plaisance; on the far side of that viaduct, the avenue's atmosphere shifted abruptly again, with a "German Village" and its adjacent "Javanese Village".

The clearest photograph of the Zoopraxographical Hall's exterior is one accidentally taken by Joseph E. Hartmann, who was engaged in capturing the fashions of Midway Plaisance strollers. Alongside the Exposition's official photographers, such as Charles Dudley Arnold, who focused especially on the main site's palatial buildings, numerous Chicago-based enthusiast photographers extensively photographed the Exposition either with a view to selling prints of their photographs or else for their own pleasure; eventually, many of their prints were donated to the Chicago History Museum. Hartmann's Midway Plaisance photographs date from May or the first half of June 1893, since Muybridge no longer operated the Zoopraxographical Hall after that period. Hartmann was intent on photographing the Midway Plaisance's corporeality. As well as focusing his attention on especially fashionable young visitors as they walked along the Midway Plaisance, Hartmann also took portraits of the concessions' managers, whose wily, unshaven faces exude profoundly disabused nonchalance and avarice as they stare into the camera, dressed in tight black suits, restless and unwillingly paused for a moment in their work; those figures (all male) appear in transit, engaged in operating their concessions for six months in Chicago before moving on to further money-generating opportunities in other countries. Most Midway Plaisance concession managers were from Italy and Germany (Muybridge was the only British "operator" of a concession, and very few were managed by Americans).

In Hartmann's photograph, the Zoopraxographical Hall is seized in its entirety, though Hartmann evidently had no interest in photographing it at all, other than as background context for his strollers' walk along the Midway Plaisance. Those young strollers, around twenty years of age, have already passed the entrance to the Zoopraxographical Hall, heading west along the avenue over the untarmaced walkway of dust, dirt and sand. They are non-spectators of moving image projection whose bodies have not been enticed inside Muybridge's auditorium, and whose senses have not been aurally beguiled by his voice's commentary or ocularly enmeshed by his sequences' propulsive visual momentum. They are oblivious to Muybridge's work, already beyond moving images even at the moment of their instigation. (By contrast, two other figures – both young and female, one dressed all in black and the other dressed all in white, the figure in white carrying an umbrella against the heat – in greater proximity to the Zoopraxographical Hall's entrance, are examining Muybridge's placards and appear to be deliberating whether to pay their twenty-five cents entrance-fee and enter into the unknown zone of moving images.) The fashionable young man wears a loose-fitting white suit and is having to carry his heavy woollen coat because of the heat; he is relaxed but his right arm is outstretched to direct his strolling movement forwards, towards his preferred attractions at the Midway Plaisance's western end. He is gazing directly at Hartmann's camera, with some degree of hostility or arrogance; he does not want to be photographed. His companion is either disinterested in the camera, or has not seen it; her tilted face appears intently focused and she is also fashionably dressed, in a long white dress and a darker waistcoat, and she is holding the young man's left arm with which he is also holding his unwanted coat. Her hat is covered with extravagant flowers and her lips appear very dark; she is gripping an unopened umbrella or large fan in her left hand. She is also relaxed but is propelling herself forward with both legs in a forceful stride – the horizontally patterned lower edge of her dress only a minimal interval above the dusty ground – so that together, those two figures appear a striking amalgam of corporeality in movement that would certainly have engaged Muybridge's eye, if he had not been inside the Zoopraxographical Hall directly behind them, conducting one of his 300 or more planned vocal commentaries on his moving-image projections.

In the centre of the Zoopraxographical Hall's entranceway, the large photographic portrait of Muybridge's face is visible; the auditorium is assigned by that portrait as belonging solely to Muybridge, indicating that he is the instigator, vocal commentator and rightful projectionist of his own moving-image sequences. Two wooden placards, leaning against the facade's central Roman columns, provide enticing details of the projections; a poster alongside Muybridge's portrait gives the cost of admission, and detachable handbills, affixed to one of the columns, impart information on Muybridge's experiments (though even with a magnifying-glass, their content cannot be read). At either side of the entranceway, two kiosks for selling tickets have been installed (in other photographs, the kiosks have awnings, to protect the sellers and spectators from heat or rain). At the instant of Hartmann's photograph, nobody is buying tickets, and the left-hand kiosk is obscured by the strolling couple. Above each kiosk, the venue's alternative name – "Zoopraxographical Circus" – is announced on hand-painted, horizontally-oriented banners. Muybridge's ticket-sellers also sold the souvenirs which are already mentioned in his Exposition contract of October 1892 (as "merchandise"): white card fans with images taken from the spinning edges of his projection discs, and replicas of the discs themselves. Few were sold, and in the following year Muybridge carried that now-obsolete merchandise back to England with him, so that those many hundreds of Exposition souvenirs now form part of his archive at the Kingston Museum, still in pristine, unfaded condition. With excessive optimism, Muybridge's architectural design had also incorporated two additional, "overspill" ticket-booths at each side of his Zoopraxographical Hall, with accompanying extra doorways, in the expectation of spectators pouring into his auditorium in immense streams; one of those unused ticket-booths, with metallic ornamentation above it, is also visible in Hartmann's photograph.

Other photographic prints of the Zoopraxographical Hall in the Chicago History Museum archive show only fragments of it, especially its left side, in images taken by amateur photographers who were attempting to photograph the more visually arresting Persian Concession (or "Persian Temple") directly alongside it but, through inept focusing, accidentally seized part of the Zoopraxographical Hall too. Among those photographs, one shows Muybridge's auditorium while it was still under-construction, with Wing's workmen installing the overspill ticket-booth at its left-hand side and completing the surrounding brickwork. Other photographers of the Midway Plaisance's figures recorded images of the relaxed occupants of moveable seats positioned around the edge of the grassgrown vacant lot alongside the Zoopraxographical Hall, for the benefit of exhausted visitors; several visitors have nonchalantly re-positioned their chairs in front of Muybridge's shaded facade, and are gazing out over the Midway Plaisance, in the exactly reverse direction to the spectators inside the auditorium, gazing at Muybridge's projection screen.

The Ferris wheel located to the west of the Zoopraxographical Hall belatedly began its operations at the end of May, a month after Muybridge's projections had begun and only two weeks or so before they ended. Many photographers, including Arnold, ascended the Ferris wheel on its opening and photographed panoramas of the Midway Plaisance and the main-site palaces beyond it, from that elevation, including sharply angled views of the Zoopraxographical Hall's flat roof and of its placarded facade. Photographers were especially active at the Exposition's opening and in the days immediately before its end-of-October closure, when the site was densely crowded with last-ditch visitors who appear bewildered in those photographs of excess corporeality, facing in all directions and unable to orient themselves. The Midway Plaisance's walkway had become a constrictive sea of mud by then. In several of those photographs, fragments of the facade of Muybridge's building are visible, but now it is excoriated, its markings annulled, and at that last moment, Muybridge and his projections are already long-gone, vanished entities.

At the moment when Muybridge signed his contract with the Exposition authorities in October 1892, he had anticipated that the Midway Plaisance would be an optimum, spacious environment for the construction of his Zoopraxographical Hall and for his moving-image projections there. At that time, the Midway Plaisance's conception was still that of an awry amalgam of corporeally populated ethnographical displays (a "Bazaar of Nations") with sites for spectators to experience unprecedented technological experimentation. Muybridge evidently anticipated that the emphasis in the Exposition main site's pavilions (such as the "Palace of Electricity") on imparting technologically oriented educational knowledge would also directly infiltrate the Midway Plaisance; he announced in his attraction's programme that his projections were endorsed by the USA's National Board of Education. The conception of the Midway Plaisance had been initially assigned to a prominent Harvard University anthropology professor, Frederic Ward Putnam, but Putnam's plans only coalesced slowly as the Exposition approached. Shortly before Muybridge's commitment to construct his venue on the Midway Plaisance, a well-known cultural journalist of that era, Julian Ralph, wrote an article entitled "Building our Great Fair" which specifies the topographical relationship of the Midway Plaisance to the Exposition's main site: "This Plaisance is a parkway connecting Jackson Park, where the fair is to be held, with a neighboring pleasure-ground. It has been added to Jackson Park for the purposes of the exposition, as an extension of the fairground. From end to end of the Plaisance concessions have been granted for exhibitions not appropriate to the main divisions of the fair, and thus the Plaisance has itself come to form one of these divisions."[7] That account emphasises that the Midway Plaisance (which most visitors entered from its eastern side, via a separate turnstiles entrance from the main Exposition site, requiring an additional payment of fifty cents), in constituting a separated "division" of the Exposition in the form of an expansive mutant entity, was characterised at its origins by its inappropriateness and its exposure of multiple spatial fractures.

That initial sense of the Midway Plaisance's tangible separation from what Muybridge must have perceived as the dignified aura of the main site only intensified as the Exposition's opening approached. The Exposition authorities hired a young, adept fairground entrepreneur, Solomon Bloom, to finalise the spatial design of the Midway Plaisance, and to accelerate and accentuate its money-generating potential. The Midway Plaisance could only exist successfully if a sufficient number of entrepreneurs were financially committed to exhibiting their attractions there, and the remaining availability of a site for Muybridge's projections in October 1892 (and the enduring emptiness of the site immediately to the east of the Zoopraxographical Hall) indicates the potential monetary emergency which the Midway Plaisance's authorities were facing, six months before the Exposition's opening. Many initial plans for social spaces along the Midway Plaisance, such as a large swimming-pool and a nursery, were abandoned and unbuilt under Bloom's direction.

A profound mismatch is apparent at the Exposition's opening in May 1893 between Muybridge's expectations of finally being able spatially to demonstrate his arts-focused and technological experimentation in moving-image projection to receptive, large-scale US and international audiences, and the actual operation of the Midway Plaisance. The rapport between the Exposition's main site and the Midway Plaisance rapidly became one of entrenched separation; visitors to the main site veered away to the Midway Plaisance primarily as an escape zone from the main site's earnest preoccupations, and for access to illicit sensorial experiences. The Midway Plaisance was largely exempt from the aura of acute surveillance imposed upon visitors to the main site, which possessed its own private police force, constantly patrolling the pavilion's interiors and the walkways between them; the Exposition's official guide describes it as: "A semi-military organisation, under the control and direction of the Exposition Company."[8] The Exposition's emphasis on intensive surveillance of its visitors would inspire future World Expositions' directors, such as those of the 1970 Osaka Exposition. On the Midway Plaisance, by contrast, the predominantly young crowds (such as the fashionable couple photographed by Hartmann) could relax or go haywire as they scanned the attractions' facades for sensorial stimuli. Before arriving at the Zoopraxographical Hall's location, visitors heading east to west along the Midway Plaisance encountered a vast hoarding above the entrance to the "Congress of Beauties" pavilion, announcing the display of "40 Ladies from 40 Nations". From behind the Zoopraxographical Hall, the Street in Cairo concession (which proved to be the most lucrative of all of the Midway Plaisance's attractions) emanated a cacophony of cries and wild music as well as the smells of Egyptian food, spices and smoke. Directly across the Midway Plaisance walkway from the Zoopraxographical Hall, the Moorish Palace enticed visitors with the sights of lightly-clad dancers performing in its interior, and had large café tables situated under awnings directly in front of it, from which seated visitors must have gazed with disinterest across at Muybridge's Zoopraxographical Hall.

The end of Muybridge's itinerance as a moving-image projectionist rendered him a static presence as an exhibitor alongside the forty other short-term proprietors of Midway Plaisance attractions. As well as in Hartmann's photographs of hard-bitten Midway Plaisance exhibitors, most of them were systematically photographed – though not Muybridge himself, who must have refused to be photographed, and took no extant photographs himself of the Exposition grounds – in on-site facial portraits by another Chicago-based amateur photographer, B. L. Snow. As with Hartmann's portraits, the faces in Snow's images appear irascible, exasperated and sardonic. Relations between those profit-driven exhibitors were abrasive and combative, mediated by many fights and quarrels. By 1893, Muybridge's experiences as a transient moving-image projectionist, alongside his other ventures of the previous decades, had made him an adept, agile entrepreneur within arts or technological environments, but on the Midway Plaisance, he appears to have been misplaced alongside his disabused, money-avid colleagues.

Muybridge's Zoopraxographical Hall is almost entirely absent from the press coverage of the Midway Plaisance in its many forms, from newspaper accounts to specialist publications such as those intended for young visitors. A large-format magazine, *The Youth's Companion*, published in Boston, issued a special Exposition edition soon after the opening of the Midway Plaisance. It surveys the Midway Plaisance attractions with caustic disdain, viewing them as forming a maleficent, crass organ appended to the Exposition's main site, and omits any mention of Muybridge's auditorium; the publication recommends that visitors avoid the Midway Plaisance entirely, and devote at least a week to assiduous visits to the main site's educational pavilions.

A prominent Chicago newspaper, *The Standard* (a self-styled "Religious and Family Newspaper" affiliated with the Baptist Church), covered the entire six-month span of the Exposition in a special weekly edition and surveyed most of the Midway Plaisance attractions over that time, but never once mentions the Zoopraxographical Hall, that absence indicating the profound negligence and oversight which Muybridge's experimental moving-image auditorium experienced. The Midway Plaisance coverage in *The Standard* begins on 18 May (two weeks or so after the Exposition's opening) with a sense of mystery and corporeally focused intrigue, as though the journalist has not so far taken the trouble to actually visit the site: "I have heard many expressions, by the way, to the paradoxical effect that the best things in the Fair are outside of it – that is, in the Plaisance. Certainly, here is the most interesting place for the study of races." By the following week, the newspaper's coverage begins to make recommendations for which Midway Plaisance concessions its readers should visit, emphasising the German Village and other "cosmopolitan" sites, and concluding: "The place [Midway Plaisance] is full of instructive and entertaining concessions." The coverage also notes that many of the concessions have not yet opened, including the immense Ferris wheel (located close to the Zoopraxographical Hall), which was installed behind schedule and then needed to be tested extensively for safety. By the beginning of June, the assessment by *The Standard* of the Midway Plaisance is reversed from its initial intrigue, and has become scathing, with the intention to discourage its readers from spending excessive time within that fraudulent, irreligious zone: "Fully one half of the performances on the Plaisance deserve to be called 'fakes', though the term is rather harsh... If you have two weeks to see the Fair, one day of it will be ample for Midway." The coverage extends beyond the mid-June point from which Muybridge's Zoopraxographical Hall had ceased its operations; by the end of June, the now fully-operational Midway Plaisance is thriving, with its summer evening weekend opening hours extended from 7pm to 10pm, and with cut-price admission to the Street in Cairo (down to ten cents) designed to pack that already-lucrative attraction beyond capacity. By the beginning of September, the Midway Plaisance was winding down, placing a mad pressure on exhibitors to foist their merchandise on gullible visitors before it became obsolete: "Everybody is eager to sell everything." The Midway Plaisance walkway was now deep with mud after summer rains, and visitors struggled through it. The last days of the Exposition saw immense, jostling crowds. The Midway Plaisance coverage by *The Standard* ends with a sense of moral outrage at its corporeal spectacles, accentuated by the entire Exposition's openings on Sundays and the wide-scale religious offence this caused: "When Chicago gives its next World's Fair two things may be predicted with some certainty – there will be no Sunday opening and no Midway Plaisance."[9]

Attendance at the Zoopraxographical Hall was certainly impaired by factors beyond Muybridge's control, especially the slow and chaotic start to the Midway Plaisance's operation; some concessions, such as the Javanese Village, only opened in July after the Zoopraxographical Hall had already closed. The opening of the Ferris wheel, to the west of the Zoopraxographical Hall, would have begun to propel far greater crowds directly past Muybridge's auditorium during the two weeks' duration when both were simultaneously operational. During the Midway Plaisance's first month or so of operation, many of the pavilions around Muybridge's auditorium had been completed, but were empty; others were still being built. That disordered start to the Midway Plaisance's operations was far from unique in the subsequent histories of World Expositions, and would be repeated, for example, at the Osaka Exposition in 1970, especially in the context of the "Expo Land" zone of entrepreneurs' concessions – beyond the main site of "official" pavilions – that exactly replicated the form of the Midway Plaisance in its role as an annex to the Chicago Exposition. Muybridge's lack of visitors and near-total absence of press coverage (the opposite of the saturation media coverage generated during his 1891 Central European projection tour through his intensive press-release approach) immediately placed the Zoopraxographical Hall under financial pressure. By mid-June, the Midway Plaisance was almost entirely operational, but Muybridge's Zoopraxographical Hall was about to disappear.

The exterior facade of the Zoopraxographical Hall can be extensively deciphered from its numerous photographic recordings. Its architecture can be traced to Muybridge's own cultural and spectatorial preoccupations, and the building's execution by Wing may be seen in the context of his extant commissions of that era, while its street-facing placards and banners can be read via magnifying-glasses or digital enlargement to verify the projections' starting times and to ascertain Muybridge's alternative name for his first-ever public auditorium constructed solely for the projection of moving-image sequences. The Zoopraxographical Hall's status as a building positioned within and against the encompassing extravagant metropolis of the World's Columbian Exposition can also be sieved from that multiply inscribed exterior carapace, as though it were itself an intricate, dense screen, doubling the projection screen within its interior. But, by contrast with the Zoopraxographical Hall's exterior, nothing whatsoever is known with certainty about that interior, in which Muybridge's spectators were confronted with his projections and his vocal commentary on them. The interior is a total mystery, paralleling all of the mysteries ever attached to moving-image projection as a sensory phenomenon, and more widely to the projectionist-inhabited interiors of cinemas as zones of obsession, compulsion, death, ecstasy, hallucination and dreaming.

By juxtaposing Muybridge's unknown interior space at the Chicago Exposition with those of the known, often-surviving auditoria of his 1891 Central European tour, some conjectural parameters for that interior emerge. Muybridge often vaunted the size of the screen on which his projections were shown, as he did in his press-release for his first Vienna event, and press accounts of that event duly confirmed that the screen held dimensions which enabled Muybridge's projections to be viewed with great clarity and detail, including in their coloured elements. Consequently, it appears highly likely that Muybridge used a large-scale screen in his Zoopraxographical Hall – with equivalent or greater dimensions to those installed in his Berlin, Munich and Vienna venues – and positioned it directly opposite the entrance, with the Zoopraxiscope projector located towards the rear of the auditorium.

Since the installation of rows of seating would have been prohibitively expensive for an auditorium designed to last only for six months before its destruction, it appears likely that the seating for spectators at the Zoopraxographical Hall consisted of freestanding, moveable chairs, so that its appearance may have resembled his 1891 Berlin Urania venue in its contemporary form, with the rows of lavish seating removed and replaced by freestanding chairs provided for the now-abandoned auditorium's last group of police-cadet spectators. It appears unlikely that Muybridge provided no seating at all (as was the case with many early cinemas of the next decade, in which audiences often stood to watch films), since all of his other auditoria entailed the seating of spectators to relax their bodies and intensify their sensorial rapport with the moving-image sequences to which they were subjected. Numerous future World Exposition auditoria for experimental moving-image projection, such as those at the 1970 Osaka Exposition, habitually required their audiences to stand, often leaning against metal bars or supports, in large part so that capacity audiences could be rapidly installed and cleared at the beginning and end of each repeated projection. 360-degree projections, such as those at the Osaka Exposition, also compelled spectators to revolve their entire bodies as well as their heads and eyes, to follow filmic actions projected around immersive screens, and therefore required them to stand upright. But since Muybridge appears certainly to have used fixed-screen projection (his Zoopraxiscope's technical capacities could not have generated immersive projection), his forward-facing, seated spectators would have focused optimally on the screen's static location, as with cinema spectators from the 1910s onwards. Standing-room auditoria at future Expositions were often designed for experimental projections lasting for shorter durations (twenty minutes, or less) than Muybridge's Exposition projections; the placard signage positioned outside the Zoopraxographical Hall, once its photographic traces have been enlarged and made legible, indicates that Muybridge was giving his projection lectures at hourly intervals, so he must have arranged his moving-image sequences, in juxtaposition with his vocal commentary, to extend for a fixed duration, perhaps for fifty minutes or so in order to allow the intermediate changeover of audiences before the start of the next lecture-projection. A projection of fifty minutes would have been at the lowest end of the durations of Muybridge's projections: his 1891 European-tour's Technical High School event in Vienna lasted for only seventy-five minutes, but Muybridge's one-off projections, such as those he undertook intermittently at the San Francisco Art Association, without tightly scheduled "cut-off" points, more habitually extended over two and even three hours.

Nothing at all is known about the assistants hired by Muybridge for his Zoopraxographical Hall projections. Their names have entirely fallen out of moving-image history. Certainly, Muybridge could not serve as his own projectionist while also standing by his screen to give his lecture commentary on his moving-image sequences. As with his employment of Ernest Webster as his surrogate projectionist for his 1889–90 tour of the UK and Ireland, Muybridge would have placed stringent corporeal demands upon his Zoopraxographical Hall operators, especially in terms of the agile speed required in changing the Zoopraxiscope's projection discs and in the operators' alert capacity to follow his vocal prompts. Those projectionists would have needed to be ready to work long hours, since the Zoopraxographical Hall was open from 8am to 7pm, seven days a week, and so it appears likely that Muybridge hired local, Chicago-based assistants.

By contrast with the multiple mysteries of the Zoopraxographical Hall's unphotographed interior, it is known for certain which sequences Muybridge projected there, since he asserted both in his Exposition programme and subsequently, after his return to the UK, that he had projected a set of around forty new twelve-inch diameter discs, prepared especially for the Exposition in collaboration with the Philadelphia-based artist Erwin S. Faber. Those discs' images were mainly based on the corporeal sequences he had photographed several years earlier during his multiple-camera experiments at his Philadelphia studio and at that city's zoo. The images were transposed to the Chicago discs' edges, as with Muybridge's previous generation of sixteen-inch discs, before being hand-coloured and enhanced. Since those twelve-inch discs still exist in Muybridge's personal archive at the Kingston Museum, their contents, projected to the Zoopraxographical Hall spectators, are certain: sequences of lightly clothed female dancers (selected from many other sequences in which the models are often naked), walking figures, figures performing rapid and erratic gestures such as abrupt turns, animals in movement and birds in flight, and riders on horses. A primary focus of Muybridge's images was in dance and performative actions, and in their projection on the Midway Plaisance avenue, those manifestations of dance must have formed a volatile amalgam – of moving-image media alongside raw, immediate corporeal dance-proximity – with the female and male dancing bodies of the adjacent Street in Cairo and Moorish Palace. The Zoopraxiscope had been adapted to project both twelve-inch and sixteen-inch discs, and it appears likely that Muybridge may additionally have shown some of his earlier discs' sequences, of horses in motion, and of his still-image experiments, such as those of urban environments. Since the merchandise Muybridge produced for sale at the Zoopraxographical Hall's kiosks was substantially unsold and was transported by Muybridge back to England in the following year, along with his new projection discs, the form of that merchandise (fans, discs) is also certain.

Subsequently, in 1899, when he had ceased all of his activity in projection, Muybridge took a negative view of the twelve-inch discs he had created for his projections at the Zoopraxographical Hall. That negativity is certainly bound up with Muybridge's calamitous experiences of mid-June 1893 in Chicago, as well as with his determined focus in the final years of his life on streamlining his conception of his life's work solely as art (rather than as commercially oriented entertainment, or as technological experimentation, since it had failed or been surpassed in both areas by that point). A second set of the discs had remained in the USA, at a New York photographic agency and storage company. Muybridge wrote from England to his collaborator in the making of the discs, six years earlier, Faber, instructing him to obliterate the Chicago discs: "These [discs] I would like to be so utterly destroyed, that no remnants of them will remain... I do not think it advisable that they should be any longer existent. If you will be going on to New York some time this summer will you do me the favor of calling at Colts' place and personally smash them into very small pieces... I now much regret having made them."[10] But Muybridge simultaneously wanted those projection discs to be obliterated, and also to survive for the future (alongside the scrapbook which distilled his life's work, and his projector and other discs). It is not known if Faber went ahead and smashed the set of New York discs as instructed, or if they were simply abandoned and eventually lost. But the 12-inch discs projected at the Zoopraxographical Hall survived, in Muybridge's own collection, because he kept his own set of them intact, and did not destroy them.

In mid-June on the Midway Plaisance, after around six weeks of operating the Zoopraxographical Hall from the Exposition's 1 May opening, Muybridge faced a catastrophic situation for his projection venue and the sudden annulling of his own status as a fixed-location moving-image projectionist. During those six weeks in which many of the surrounding pavilions remained unopened and when visitors to the Midway Plaisance often went no further west than the Woodlawn Avenue viaduct, the Zoopraxographical Hall had only a scattering of spectators who entered its interior to experience Muybridge's projections and accompanying vocal commentary. Most visitors who did go that far advanced immediately to the Street in Cairo attraction behind Muybridge's venue, oblivious to its facade's enticements and its banners' promise of viewing a form of "circus", or else installed themselves in the spacious café areas outside the Moorish Palace on the Midway Plaisance's far side. The Zoopraxographical Hall's costs and wages accumulated far in excess of its takings, and Muybridge – who appears to have been paying for everything personally, without support from the University of Pennsylvania or any of his other past sponsors – was confronted with imminent financial ruination. Muybridge's calamity also formed an initiating disaster for moving-image projection and its temporary conjuring-away as a spatially installed entity. Any prospective moving-image exhibitors or entrepreneurs aware of Muybridge's dearth of spectators would have been deterred from envisaging the instigation of their own spatial environments, and the era exclusively of the itinerant moving-image projectionist extended for a further decade or more until the eventual mass-construction of urban cinemas from around 1907-08. Muybridge's 1893 innovation in undertaking his projections in a specially designed environment was damned at its first moment as a spectral, diabolical conception. Muybridge was about to lose his Zoopraxographical Hall to another exhibitor, who had no intention to inhabit that space with moving images.

On 16 June, the *Chicago Tribune* reported that a "procession" had taken place on the Midway Plaisance several days earlier, and Muybridge was mentioned by name – as "Professor Muybridge" – as one of its notable participants. All of the Midway Plaisance's exhibitors, together with their venues' staff, performers and animals, had formed an insurgent parade that then traversed the turnstiles boundary between the Midway Plaisance and infiltrated the Exposition's sedate main site – music blaring, and outlandish, alien bodies exposed, as though in a sudden revolt of fairground freaks, belly-dancers, hoodlums and misfits, directed against the Exposition's staid tourists and educationalist figures of power – to circuit the main site's palaces under the severe eyes of its "semi-military" police, before returning to their attractions.[11] Whether that procession was intended to publicise the financially ailing Midway Plaisance concessions to Exposition visitors otherwise untempted to cross over to its more volatile annex, or else to form a momentary insurrection, is not clear from that one cryptic account. Muybridge's decision to leave behind his work at the Zoopraxographical Hall to participate in the procession, thereby overturning the schedule inscribed on his placards, and breaking apart his rigid solitude to ally himself corporeally with the other Midway Plaisance exhibitors, must have formed a unique, last-ditch act. It would have been apparent to him by that point that the Zoopraxographical Hall could not survive for the entire Exposition, which still had over four months to run. And that procession's exposure of his distinctive, tall figure – with his long white hair and beard, and black suit – must have cast him into a profoundly inhabitual space of vulnerability, opened-up to approaches from his abrasive fellow-exhibitors.

Within a day or two of that procession, the Zoopraxographical Hall was abruptly closed-down, with Muybridge's projection auditorium adroitly manoeuvred out of his hands. It almost immediately re-opened, its mysterious interior space now usurped by another attraction, the "Pompeii Panorama". Another panorama building already existed on the Midway Plaisance to the west of the Zoopraxographical Hall, immersing spectators in concave painted panels of the contemporary eruptions of the Kilauea shield volcano on the island of Hawaii (continuous eruptions had taken place in the preceding years, though they would temporarily cease in 1894); that panorama pavilion had been specially built as a large, high and near-circular construction with vast exclamations of enticing text imprinted upon it and with an ornate ticket-booth in front, surmounted by a statue of Pele, the Hawaiian female fire-deity. By contrast, the oblong, relatively low building which Muybridge had designed for his projections and commissioned Wing to build was acutely ill-suited for the installation of a panorama. Even so, the Pompeii Panorama took the pre-eminent position for any spectators interested in experiencing panoramas, or representations of natural catastrophes, since almost all visitors travelled east to west along the Midway Plaisance and would have encountered the Pompeii Panorama before reaching the "Volcano of Kilauea Panorama"; the historical aura of the Pompeii Panorama may also have enticed visitors already imbued with the historical emphasis of many of the Exposition's main-site pavilions. Despite the mis-aligned dimensions of Muybridge's hall for the exhibition of a panorama, the Pompeii Panorama proved financially viable, and probably profitable. The charts of financial documentation eventually compiled by the Exposition authorities for all of the Midway Plaisance attractions, and conserved as copies at the Chicago History Museum, show that the Pompeii Panorama took $19,500 in admission charges (with one third of that swallowed by the Exposition authorities, in addition to operational costs) over its four months of existence, ranking 31st among the 41 concession operators.

That sum was proportionally far greater than Muybridge's takings. He took an astonishingly low sum of $320 from six weeks of operation. He was ranked in the 41st of 41 places by a wide margin, and even attractions that were also forced to close prematurely made far more money. The extreme innovation of the Zoopraxographical Hall is intimately doubled by the sheer distinctiveness of that excessive non-profitability, as though Muybridge, with his intractably perverse obstinacy, had somehow intentionally sabotaged his own venue's financial regime (and that of future cinema spectatorship). Muybridge may have sometimes operated the Zoopraxographical Hall without an admission charge, and depended on merchandise and programme sales (the Midway Plaisance clientele could certainly not have afforded copies of his immense *Animal Locomotion* publication), especially when it became clear that visitors needed demonstrably to be seen entering his auditorium.

But photographs of the Zoopraxographical Hall's facade show clearly that he was usually charging twenty-five cents (as the Pompeii Panorama, and most other Midway Plaisance attractions, also did). His total fee-paying audience amounts to 1,280 spectators, around thirty each day, if he was operating the Zoopraxographical Hall for seven days a week over six weeks or so, as appears highly likely, since he rarely took holidays (the last one being his own ill-omened visit to Pompeii and ascent of the Mount Vesuvius volcano, in 1891, during his projection tour of Europe).

The Pompeii Panorama was an Italian (or Italian-American) operated attraction, run by the exhibitor Gennaro Pandolfelli. In the Chicago History Museum archives, a copy of a handwritten memorandum of the two exhibitors' agreement, ostensibly intended for the attention of the Exposition authorities and appended to a copy of Muybridge's October 1892 contract, mentions a recent meeting between Pandolfelli and Muybridge in which the building was assigned for the remaining duration of the Exposition to Pandolfelli, for an unspecified sum (no trace exists of a contract between Muybridge and Pandolfelli, if any contract ever existed at all). Pandolfelli appears not to have operated other Midway Plaisance concessions and, as with Muybridge, makes no appearance in Snow's series of photographic portraits of exhibitors, unless he used other names or aliases. Muybridge's loss of his auditorium appears an opportunistic appropriation of it by an alert scavenger watching out for financial failures. It may be that Pandolfelli seized the chance to approach Muybridge during the "procession" from the Midway Plaisance to the Exposition's main site. There may also have been some complicity between Pandolfelli and the Exposition authorities, who must have been actively gratified to see the Zoopraxographical Hall's closure, from which their total cut was only $100 or so, and its supplanting by a more lucrative attraction.

Muybridge, in his later writings and documents such as those incorporated into his scrapbook of 1894-1904, never once mentions his loss of the Zoopraxographical Hall, and only refers to the Chicago Exposition at all in the context of its prestigious status and of his having lectured there, rather than his own shortlived involvement in it as a Midway Plaisance exhibitor, about which he remained utterly silent for the rest of his life. All that can be ascertained from that silence – in its contrast to Muybridge's insistent repetition of what he considered to be his work's triumphs – is that his encounter with Pandolfelli and the slipping from his hands of his Zoopraxographical Hall formed deeply negative events for him.

The opening of the Pompeii Panorama was rapidly announced; a new edition of the *Official Guide to the World's Columbian Exposition*, printed in mid-June, already registers the voiding of the Zoopraxographical Hall. Most visitors to the Midway Plaisance bought and carried the regularly updated *Guide*, and the rapidly executed change in the Zoopraxographical Hall's status must have been included as an "update" at a last-minute stage. The *Guide* omits the Zoopraxographical Hall and lists in its place the "very realistic" Pompeii Panorama, painted by "Count" Antonio Coppola, who specialised in representations of the eruption of Mount Vesuvius. Coppola was present at the Exposition and, like Pandolfelli, was from Naples (Coppola and Pandolfelli conceivably made Muybridge "an offer he can't refuse" in order to appropriate his auditorium, though they may also have been wary of Muybridge, the black-suited killer: a self-confessed, unpunished assassin, who had taken his gun and coolly shot dead his wife's lover with one well-aimed bullet to the heart.) The *Guide* asserts that the panorama had demanded "three years of incessant labour" for Coppola to complete it, though in reproduction, the painted panorama appears ineptly and rapidly rendered, showing a maladroit combining of the 1890s Pompeii ruins and the city in the course of its AD79 volcanic-ash immersal. The *Guide* summarises the new attraction as: "Showing the ancient city, eruption of Vesuvius, modern Pompeii, etc; admission 25 cents."[12]

Naples itself already possessed a long history of the exhibition of panoramas showing Pompeii's destruction (a genre known as "Pompeiorama"), and some of Coppola's other smaller-scale renditions of his subject – which he continued to paint over the following two decades – are conserved at the Naples Archaeological Museum, though the large-scale panorama exhibited in Chicago appears not to have survived, along with most of the other artworks displayed at the Exposition (and those exhibited at future World Expositions), obsolete and consigned to oblivion at its closure, and burned. Coppola must have brought that panorama's panels with him to Chicago with the determination to find a belated chance to exhibit it, since its installation in Muybridge's auditorium was so rapid and abrupt (it could not have been shipped from Naples in the interval between Muybridge's expulsion and the panorama's opening); the horizontally oriented scale of its panels enabled them to be installed in circular form in Muybridge's building with its relatively low ceiling. It occupied Muybridge's building, as what must have appeared to him as an intruding, contaminating presence. From the few photographs showing the building (now devoid of its "Zoopraxographical Hall" title above the entrance) that were taken after mid-June, it appears that its new operator Pandolfelli did not take the trouble to position placards, posters or banners outside, as Muybridge had done, or otherwise alert passers-by to what was inside, evidently relying on the Midway Plaisance's *Guide* to direct visitors into his attraction.

The photographic historian Julie K. Brown, in her 1994 book *Contesting Images: Photography and the World's Columbian Exposition*, offered some barbed criticism of Muybridge as a Midway Plaisance operator, proposing that he should himself have presented a panorama of calamity rather than focusing on his unprecedented moving-image projection experimentation: "Clearly, exhibitors like Muybridge had something to learn from the promotion of the recent volcanic events, dramas of history, and landscape views that were successfully packaged on the Midway for the visual entertainment of spectators."[13] Muybridge had already experimented with the photographic panorama format around fifteen years earlier, with his two urban panoramas of San Francisco of 1877-78, and had exhibited his panorama work in an immersive format, surrounding and disorientating its spectators, at the Art Association exhibition rooms in San Francisco. He evidently remained enduringly attached to his work in the panorama medium since he transported a complete set of one of the large-scale panoramas back to Kingston-on-Thames in the following year, but he had envisioned an entirely disparate, experimental project for the Exposition, with his moving-image projection auditorium.

To evacuate the Zoopraxographical Hall, Muybridge would have needed only to transport his Zoopraxiscope projector, discs and merchandise to his Exposition boarding-house room alongside Lake Michigan. As with the Zoopraxographical Hall, most of the other Midway Plaisance attractions contained tenuous, momentary sensory conjurations, of dancing bodies or deliquescing images, that left nothing behind, and the monumental painted panels of the Pompeii Panorama formed an aberration in that regime of imminent disappearance. But Muybridge stayed-put in Chicago for several more months. His occupation as a Midway Plaisance entrepreneurial exhibitor was expunged and his status now returned to that of a once-eminent innovator in photographic art whose work was sliding out of its past prominence. He now solely inhabited the Exposition's palatial main site, albeit as a marginal presence. He undertook a badly-attended projection event on the evening of 31 July at the Exposition's Congress on Painting and Sculpture – one of its many "Congresses" spanning religion, technology, art, architecture and other subjects – in a lecture-room at the Art Institute of Chicago, the Exposition's art-focused venue in the centre of the city (while its Palace of Fine Arts occupied a primary position in the Exposition's main site). At the Congress event, Muybridge, with his pivotal obsession with repetition, projected exactly the same moving-image sequences, and gave the identical vocal commentary, as he had in his Zoopraxographical Hall, thereby erasing the division between the Exposition's peripheral and official zones, as had already occurred with his participation in the previous month's infiltrating procession of Midway Plaisance exhibitors into the main site. With that projection at the Art Institute of Chicago, Muybridge had returned to his prior status as an itinerant moving-image projectionist, his experiment in fixed-location projection permanently curtailed. A brief account of his 31 July projection was included in a *Chicago Tribune* article of 3 August that summed-up unpopular public events at the Exposition, headlined "TALK TO EMPTY SEATS". The article wryly commented that at Muybridge's event, "The attendance was not large".[14]

After the Exposition had ended, its immediate aftermath became that of a future-oriented genealogy of incinerations, alongside demolitions and jettisonings of detrita into Lake Michigan. Many of its now-unwanted buildings were destroyed intentionally by fire, others in accidental fires; the Chicago History Museum archives contain innumerable photographs of burning or burned-out pavilions, as though those buildings' incinerations had fascinated the city's many amateur photographers even more than their extant, palatial forms. An accidental fire, destroying the Cold Storage building, had taken place even during the Exposition's period of opening, on 10 July 1893. The main fires began at the beginning of the following year. An initiating accidental fire of 8 January 1894 destroyed the Exposition's Peristyle, Casino and Music Hall buildings; the Agricultural Building was intentionally destroyed, on 7 February 1894, as was the Illinois State Building, on 18 February 1894, the Horticultural Building, on 26 February 1894, and the Palace of Electricity, on 5 July 1894. The immense Manufacturing and Liberal Arts Building was photographed in its reduction to a tangled, collapsed metal infrastructure, after its destruction by fire. The photographic documentation of the vacated Exposition's scorched-earth lands prefigured many other future Expositions' oblivious incinerations and abandonments. By contrast,

Chicago retained its enduring memorial urban-imagining of the majestic, dream-illuminated "white city" of the Exposition.

Once the Exposition was over and the Pompeii Panorama became obsolete, Muybridge's building returned to its status as the first-ever moving-image projection auditorium, its "Zoopraxographical Hall" insignia erased but Muybridge's canopied ticket-booths and ornamented Roman columns still intact. The Zoopraxographical Hall was eventually razed; with its brick construction, it could not have been readily incinerated, and its bricks may have been re-used, conceivably by Wing for elements of his subsequent Lange House, or its detrita may have been buried in a pit beneath the Midway Plaisance. But first, the rest of the Exposition site was obliterated, as though to save Muybridge's auditorium to last. A file of Chicago History Museum photographs titled "Demolition" contains a small, anonymous photographic print labelled "World's Fair Chicago 1893 Wrecked after Closing", which must have been taken in 1894. The Exposition site appears as an immense wasteland of striated mud and tangled debris resembling a World War One battlefield; that wasteland perhaps invokes Muybridge's own subsequent Chicago Exposition memories and wounded silence, though he evidently had an enduring attachment to his memories of Lake Michigan. Looking at that wasteland photograph closely, with a magnifying glass, it becomes clear that the vast scorched-earth terrain in the foreground is the main Exposition site and that the Midway Plaisance buildings remain intact in the left-hand-side background, including the minaret of the Street in Cairo, and, as a minor low-lying shadow in front of that larger attraction, the barely distinguishable shell of the Zoopraxographical Hall. The Midway Plaisance buildings – the most negligible and sensorially awry element of the Exposition, its annex and experimental antithesis – were destroyed only at the end.

Muybridge preferred his relatives to believe that he had returned to Kingston-on-Thames from Japan, rather than Chicago, when he eventually reappeared in his riverside place of

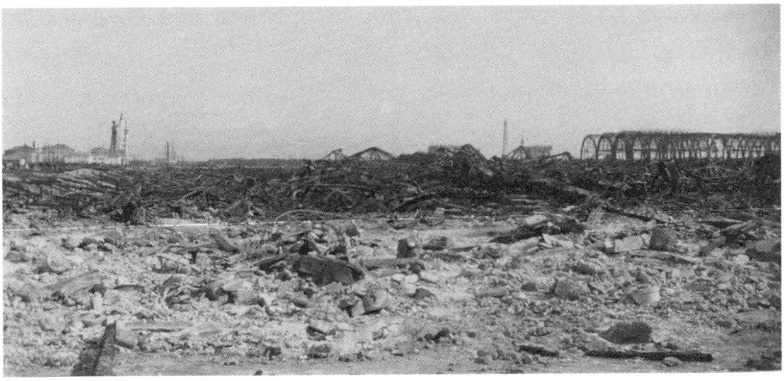

Ill. 13: "World's Fair Chicago 1893 Wrecked after Closing", 1894.

birth. That departure point in Japan cannot have appeared implausible or impossible to his relatives, since they certainly believed him; he had publicly announced his intention to undertake a projection tour to India, Australia and Japan after his 1891 Central European tour, and had planned and prepared it before eventually deciding to prioritise his Zoopraxographical Hall project for the Chicago Exposition. Muybridge's relatives would have believed that his planned global tour had been concertinaed to its Japan component. Along with his long-standing interest in Japanese art, Muybridge would have seen, and perhaps even met, some of the many visitors from Japan (educationalists as well as artists, philosophers and writers) at the Chicago Exposition, which held a large, lagoon-side wooden temple, the Ho-o-den, built by Japanese craftsmen to simulate a celebrated temple in the Kyoto region and known at the Exposition as the "Phoenix Palace", as well as tea rooms (as with other Exposition buildings, the temple and tea rooms were burned to the ground after the event's closure). Engagement with photographic culture, and with its innovations and experiments, was in its ascendancy in urban Japan in 1894. Muybridge's conjured visit to Japan in the autumn of that year now oscillates between an imaginary fabrication intended by him to divert attention from his still-recent Chicago catastrophe, a phantom but feasible projection (or planned projection) as with those he undertook in cities such as Prague during his tour of Central Europe's cities, and even an actual visit from which all verifiable traces have been lost.

None of Muybridge's immediate family was still living in Kingston-on-Thames when he returned there, but he had many female and male cousins who lived and worked in the town (one of them, George Lawrence, was a corn merchant) and he was evidently concerned to integrate himself into that extended family on his return, while initially maintaining his distance by resuming his habitual boarding-house room existence. The cousins communicated to one another by letter, and at least two extant letters from November 1894, exchanged between Muybridge's cousins Rachel Rickard and Elizabeth Selfe, mention his return in the context of a journey from Japan: "Edward Muggeridge [Muybridge's real name] has come to Kingston from Japan and is lodging in Mr Gennett's house in the High St".[15] "Mr Gennett's house", known as "The Chestnuts" boarding-house, was located on a busy road of wharves and shops directly next to the river Thames, and Muybridge lived there for around five years (apart from a trip back to the USA to resolve disputes over the ownership of his negatives). He had been born in the same street. Muybridge had a small, rear-facing room and for part of his time there, the adjacent room was occupied by a piano-tuner. He worked on his scrapbook there and inscribed the boarding-house's name on its opening page in a list he assembled of his addresses of the past three decades. He then ended his boarding-house existence and moved, in 1899, with two of his cousins and a domestic servant, to the upper storey above a shop located at 161 Kings Road in the newly constructed Canbury Estate of many hundreds of mainly terrace-houses, north of Kingston-on-Thames' centre, and finally, in 1902 and again with his cousins, to a more spacious, detached house nearby with a large, walled back-garden, at 2 Liverpool Road, close to the entrance to Richmond Park. By that time, his Chicago-decimated finances must have significantly improved with the publications in London of two volumes of his work, *Animals in Motion*, 1899, and *The Human Figure in Motion*, 1901.

The cousins' letters of November 1894 on Muybridge's arrival back in Kingston-on-Thames mention his evocation of visits to the "3 great European-style cities of Japan", without mentioning their names.[16] Those three cities may well have been the adjacent cities of Tokyo and Yokohama, and the port-city of Hakodate, on the northern island of Hokkaido, with its hillsides' gridded, sheer streets that resemble parts of the topography of San Francisco. Hakodate had been a primary destination of international visitors to Japan especially since the beginnings in 1868 of its Meiji-era opening-up to European technological, educational and artistic influence, with a large European population of entrepreneurs as well as embassies and Christian churches. Yokohama formed a similar port-environment devoted to transmitting European innovations into urban Japan, and Tokyo was the primary location of Japan's rapidly emerging photographic industries and exhibition halls.

125 years after Muybridge announced to his cousins that he had just arrived from Japan, I travelled there to attempt to discover any trace of that lost visit. That journey into Muybridge's Japan appeared at an impasse even at its origins. No photographic or urban historian of 1890s Japan had ever heard of Muybridge projecting his work there, or even of his appearing in person at any of its cities' technological or trade expositions of that era. But there existed no reason why he should not have gone there then, as other photographers and artists from Europe and the USA had. Tokyo had been destroyed twice since 1894, in a great earthquake in 1923 and then through World War Two aerial fire-bombing raids that also destroyed Yokohama. No auditoria in which Muybridge could conceivably have projected his work in 1894 in those two cities had survived. By contrast, contemporary Tokyo and Yokohama formed environments pivotally inhabited by Muybridge's seminal moving-image innovations and by his corporeal and urban obsessions with projection, in their digitally screened cityscapes.

In Hakodate, a succession of 1880s single-storey British-built red-brick warehouses still bordered the marine quaysides, now occupied by fashion emporia and coffee-shops. In the mid-1890s, still newly-built, those warehouses had been used by the city's international entrepreneurs to assess new technological (among other) inventions brought to the city by itinerant, in-transit inventors, or inventors' representatives, seeking opportunities to display or exploit those technologies to Japan's urban citizens in their avid desire for innovations. If Muybridge had really travelled to Hakodate, he could conceivably have demonstrated his work to closed groups of entrepreneurs in those warehouse environments, which were also widely employed in that era for public performances of music, magic-lantern spectacles, dance and other manifestations of new cultures. For the most startling or beguiling of those innovations, tours or demonstrations could be arranged by promoters within northern Japan's many European-style grand urban auditoria, including those of Hakodate itself. In the contemporary silence of the quayside red-brick warehouses, after their clientele had left for the night, the sonic residues of those experiments and their audiences' exclamations of awe still appeared adhered to the shadowed ceilings.

After several weeks of searching, I left Japan with no tangible or archival trace of Muybridge's 1894 presence there whatsoever. I had to conclude that Muybridge's narrative for his Kingston-on-Thames cousins had been an imaginative fabrication – an originating lie, or delirium, or delusion, of moving-image projection – concocted to enthral those gullible cousins with the exotic attractions of his travels, or else to avert any discussion of his Chicago projections' disaster. Almost always, Muybridge rigorously edited-out and obliterated any evidence of what he viewed as negative in his life or work, most systematically through the scrapbook project that occupied him for the final decade of his life. For him to abandon that habitual strategy of reduction and erasure, and instead actively elaborate a fraudulent narrative, as with his journey from Japan, as an aspirant projectionist of moving images, travelling from city to city there, appears unique.

Muybridge undoubtedly travelled back to England in 1894 from a starting-point in the USA, and not from Japan. He is verifiably included in the passengers' list for an Atlantic crossing of the Cunard liner, RMS Umbria, leaving New York's Hudson River Pier 40 at Clarkson Street and arriving in Liverpool on 8 September 1894, travelling as a "Gentleman Cabin Passenger" for the single-ticket fee of $45. Even so, Muybridge's phantom journey to Japan remains conceivable, since over the following two months he could have travelled there, in the form of an arduous overland and steamship "expedition" – in the desperation of a last-ditch determination to cancel-out his Chicago experiences with new, unprecedented successes in Japan's cities – and having investigated the potential to project his moving-image sequences in those cities and concluded that no opportunity existed for those projections, finally returned to Kingston-on-Thames at the beginning of November. The work of all projectionists is generated from obsession and adroit malleability; as the instigating moving-image projectionist, Muybridge's envisioned projects and travels, such as his madly conjured 1894 journey to Japan, are always finally viable and plausible, even when they appear demonstrably to be unhinged untruths. The traces of that journey to Japan may simply have become so multiply and thoroughly erased – through 1920s earthquake seisms, then in 1940s fire-bombed negations of memory's archival residues – that they now constitute a determining absence, pre-eminently inscribed in the motion-capture enthralled, infinitely-screened urbanscapes and voids of contemporary Tokyo.

Once he had installed himself in his boarding-house room in Kingston-on-Thames in November 1894, Muybridge irrepressibly began to plan new projections; but from that moment on, the spatial parameters of his work as a projectionist were confined to England, and he had returned to his former itinerant inhabitation of pre-existent auditoria, after his expulsion from his self-designed Zoopraxographical Hall. His final projection-tour extended intermittently over the course of six months, from October 1895 to March 1896, and as with his tour of Central Europe's cities, Muybridge publicised it with a press-release, in May 1895, on paper headed with his address: "The Chestnuts, Kingston-on-Thames". Muybridge emphasised in his press-release that he would be projecting forty new colour discs which had never previously been seen in the UK. (These were the discs whose New York-stored set he would instruct Faber to smash, four years later.) Muybridge mentions his presence in Chicago for the only time in that press-release, in the context of the active emphasis on repetition that was so vital to his work: "The present [projection event] will be virtually a repetition of the series of Lectures given... at the Columbian Exposition."[17] He asserts that he will present his projections on a large-format screen – "much larger than the size of life" – and (following his financial losses in Chicago), emphasises the size of the fee, a relatively large one in the context of that era, which he will exact for his work. That financial consideration is very rare for Muybridge, who appears almost always to have been oblivious to the monetary dimensions of his work, but his press-release's exacting stipulations about the conditions of projection in the auditoria he will visit, by contrast, are familiar, long-engrained ones. Muybridge is willing to extend the "regular" span of his tour, so that it could effectively continue indefinitely: "In the event of engagements being made sufficiently early to ensure their inclusion in the proposed regular tour, the fee for each lecture will be Ten Guineas, which will cover expenses of assistants, oxy-hydrogen, light, etc., – it is, however, expected that the Lecture-room will be put into proper order for placing the apparatus and hanging the screen... For engagements made after the regular tour has been arranged, transportation expenses from the point of departure will be charged in addition to the fee – if the invitation can be accepted."[18]

Muybridge updated that first press-release with a further one in October 1895, in which he emphasised that his new projection lectures, undertaken in itinerance across numerous UK cities and towns, such as Sheffield, Liverpool and Ipswich, had received enthusiastic responses, though the press accounts he lists of them – such as: "Perhaps the most unique lecture ever delivered in Birmingham", from the *Birmingham Post* – actually date from his 1889-90 tour with his "operator" Webster, rather than from his most recent one.[19] He had projected his work in cities' civic halls and in the assembly rooms of public schools, working once again with a hired projectionist for his Zoopraxiscope. Gradually, over the following two years, his glass-disc projection events grew further apart and rarer, as though his obstinate capacity to project were now finally extinguishing itself in decelerating, jolting exhaustion, while young inventors' celluloid projections were beginning to proliferate.

Muybridge's final projection took place in May 1897 in Cornwall, at the St Ives Arts Club, which had been established seven years earlier. He must have made an arduous journey with his projection equipment, travelling by train for many hours from Kingston-on-Thames to St Ives. The small, wood-panelled room above a carpenter's shop, holding around forty spectators and situated directly against the seafront's crashing waves, formed a highly disparate venue to the far larger, often-palatial auditoria in which Muybridge had projected his work to corporeally packed, awestruck audiences during his 1891 tour of Central European cities. That upstairs room was reached only via a precarious ladder propped against the building's exterior, and it appears that Muybridge may not have been able to use his Zoopraxiscope to demonstrate his moving-image projections; the Club's surviving bound record-book of events for that era indicates that Muybridge was restricted to using a slide-projector to show only still images. At his last projection, Muybridge would certainly have undertaken his work in a condition of white-hot fury.

Muybridge worked assiduously on his scrapbook until 1902, but interest in his work declined so sharply in the twentieth-century's first years that he had almost no material with which to update it, other than review cuttings of the two volumes he had recently published of his work. Once he had moved to his final house, in Liverpool Road, he focused his attention on the large back-garden. His first project there was to bury his negatives (which had been the focus of rancorous ownership disputes several years earlier), so deeply and successfully that most of them have never been discovered. He then formed the project of arduously digging large earth pits with the apparent aim eventually to construct a network of swimming holes or pools replicating the configuration of the US Mid-West's Great Lakes, including Lake Michigan, directly beside which he had lived in 1893 during the months of his Zoopraxographical Hall projections (and also for the months following his expulsion from his auditorium), in the Exposition's boarding-house. Muybridge may have been driven by residual affection for Lake Michigan, or else by the enduring need to erase still more comprehensively his memories of his Chicago ordeal by transmutating and supplanting it with his re-created garden topography. He was not the only creator of swimming holes designed to manifest his personal history of itinerance; during the same period, twenty miles or so to the south-west, the retired explorer of African rivers and formerly relentless traveller, Henry Morton Stanley, was also digging or adapting swimming lakes in the grounds of his Furze Hill Place mansion in Pirbright (though Stanley wisely commissioned labourers to dig them for him), and naming them after locations he had once travelled to or had "discovered", such as a swimming lake named "Stanley Pool" after a wide stretch of the Congo river. (If Muybridge had completed his back-garden project, he could perhaps have named it "Lake Muybridge".) That back-garden project also appears prescient of performance-art actions of the later twentieth century, or land-art works such as those of another artist closely preoccupied with moving-image projection, Robert Smithson.

Muybridge worked nocturnally, naked, his figure in movement, until his death on 8 May 1904 in his back-garden while immersed in that exhausting final project, which appears to have formed an excessive medium of corporeal projection, in its attempted autonomous spatial re-manifestation of Lake Michigan, thereby overruling his Exposition fixed-location moving-image projections' forcible cessation. Even working without (or beyond) moving images, a projector, negatives, auditoria, or audiences, Muybridge remained a projectionist. Death and projection are always intimately conjoined. The re-designed back-garden of 2 Liverpool Road still holds a leaf-covered abandoned swimming pool, underneath the fronds of Californian palm trees, which may have its origins in Muybridge's last excavatory experiment. In 2010-11, the artist Becky Beasley exhibited her extraordinary sculptural and photographic project *8th May 1904, Kingston* (titled after the date of Muybridge's death, and the result of intensive research into his personal archive), which involved conjoined topographical explorations of his Liverpool Road back-garden, which she photographed extensively, along with his San Francisco panoramic work and its ambitious urban mapping.[20] The garden, with Muybridge's still deeply-buried negatives under its contemporary surface, appears a still-animated spectral site of Muybridge's obsessions.

I visited the Kingston Museum to view the original Zoopraxiscope projector with which Muybridge toured Central Europe's cities in 1891, and used for his Zoopraxographical Hall moving-image projections in 1893. The museum's construction, as well as that of its adjoining public library, was funded by the industrial philanthropist, Andrew Carnegie, who attended the library's opening ceremony, which was certainly attended also by Muybridge, though he died shortly before the museum's subsequent opening. In a corridor of the museum building that remains largely unchanged since its construction, the Zoopraxiscope is kept in a glass case. A few of its parts are missing, rendering it non-functional, but with the adroit attention of a grandmaster projectionist or "operator", it could rapidly be ready to project once again: an auditorium is primed in expectation for it, on the museum's upper storey. Some of

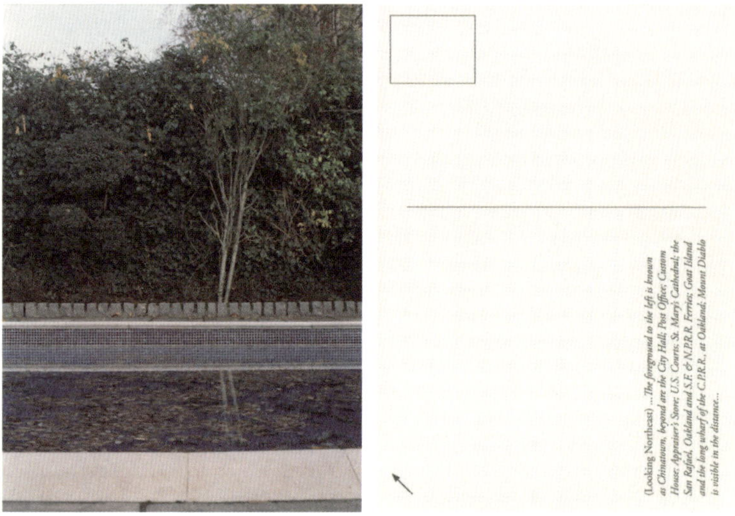

Ill. 14/15: Becky Beasley, postcard artwork (photographed in Muybridge's Kingston-on-Thames house back-garden, with swimming pool), "Looking Northeast", recto and verso, detail from *8th May 1904, Kingston*, 2010.

Muybridge's projection discs are kept nearby, and also close at hand are the reinforced travel cases in which he transported his projector and discs, their surfaces multiply scuffed and densely layered with the adhesive labels of many countries' shipping and railway companies from the Zoopraxiscope's travels, city to city, from the period of sixteen years of Muybridge's initiating corporeal inhabitation of the status of the moving-image projectionist.

Notes

1. World's Columbian Exposition Planning Committee, *Report of the Committee on Grounds and Buildings*, Chicago World's Columbian Exposition, Chicago, 1891, page 4 (document conserved in the University of Chicago Library Special Collections). This chapter was principally researched in the Library's Special Collections devoted to the Exposition, as well as in the Chicago History Museum's photographic and document collections relating to the Exposition.

2. World's Columbian Exposition Concession Agreements, volume 3, number 8 (document conserved at the Chicago History Museum).

3. Eadweard Muybridge, *Descriptive Zoopraxography*, programme for the Zoopraxographical Hall, University of Pennsylvania, Philadelphia, 1893, page 1.

4. Ibid, page 2.

5. *Chicago Tribune*, 2 March 1916, page 15.

6. *Descriptive Zoopraxography*, page 2.

7. Julian Ralph, "Building our Great Fair", in *Harper's Weekly*, New York, 17 September 1892, page 898. Ralph also wrote a book about the Exposition site and its relationship with the urban culture of Chicago, *Chicago and the World's Fair* (Harper and Brothers, New York, 1893), which was published even before the Exposition had opened.

8. *Official Guide to the World's Columbian Exposition*, edition of mid-June 1893, Chicago World's Columbian Exposition, Chicago, 1893, page 170.

9. Bound scrapbook of cuttings (dated but without page numbers) from the Exposition coverage by *The Standard* newspaper, conserved in the University of Chicago Library Special Collections.

10. Muybridge, manuscript letter, 16 June 1899, Muybridge Collection, University Archives and Records Center, University of Pennsylvania, Philadelphia.

11. *Chicago Tribune*, 16 June 1893, page 2.

12. *Official Guide to the World's Columbian Exposition*, edition of mid-June 1893, page 34.

13. Julie K. Brown, *Contesting Images: Photography and the World's Columbian Exposition*, University of Arizona Press, Tucson, 1994, pages 104-105.

14. *Chicago Tribune*, 3 August 1893, page 10.

15. Rachel Rickard and Elizabeth Selfe, photostats of exchanged letters, 6 and 12 November 1894, unlabelled Selfe-family correspondence folder, Kingston Local History Room, Kingston-on-Thames. (The "Local History Room" was located in a building of Kingston's Tiffin Girls' School, in the basement of which Muybridge's personal archive was stored for many

years by the Kingston Museum; the Local History Room relocated in 2015 to Kingston's Guildhall building as the "Kingston History Centre".) The correct name of the owner of "The Chestnuts" boarding-house was Jemmett, rather than "Gennett"; he also lived there, and worked as a solicitor as well as a boarding-house proprietor. The boarding-house, located on the opposite side of the High Street from the riverbank, was demolished in the 1930s.

16. Ibid.

17. Muybridge, personal scrapbook (1894-1904), unpublished, conserved at the Kingston Museum, Kingston-on-Thames, page 164. Muybridge inserted his 1895-97 projection tour's press-releases into his scrapbook, as a current update of his work.

18. Ibid, page 164.

19. Ibid, page 166.

20. *8th May 1904, Kingston* was exhibited at the Stanley Picker Gallery, Kingston-on-Thames, from 24 November 2010 to 5 February 2011. I'm grateful to Becky Beasley for presenting that project at a conference on Muybridge's work held at the Normansfield Theatre, Hampton Wick, in November 2016.

Projection's Devils: Max and Emil Skladanowsky, 1895-97

Muybridge's Central European moving-image projection tour of 1891, and his own corporeal figure as a public moving-image projectionist, inspired many experiments among inventors, photographers and technologists in the cities he passed through; above all, Muybridge inspired the Skladanowsky brothers, Max and Emil, with the second of his Berlin projection events at the Urania auditorium on 11 March 1891. Four years later, after the pivotal event of celluloid film-stock becoming commercially available for the first time, the Skladanowskys would project their own films in the lavish ballroom of the Central Hotel in Berlin, as the first-ever public projection of celluloid films to a paying audience, using a projector which they had designed and built themselves, the "Bioskop". Muybridge's figure at the Urania and his formative status as a moving-image projectionist – showing sequences he had created himself, and vocally commentating on them in direct contact with his spectators – activated the Skladanowskys' ambitions far more than the material form of the Zoopraxiscope itself, though they would go on to devise their own distinctive corporeal manifestation in moving-image projection's space. The Skladanowskys had few resources, but they proved to be technically adept inventors and, as well as building their own projector, they were also able to construct their own film-cameras and adapt raw Eastman-Kodak celluloid for projection, in defiance of the apparent impasses which defeated many of their competitors of that era.[1]

The Skladanowskys were projection's adroit devils: devious young rogues who, always close to poverty, were financially driven. Travelling to the elegant Urania auditorium in central Berlin from their base in the rough, overcrowded environment of the city's northern Prenzlauerberg district, with its innumerable breweries and chemical industrial pollution, they must have appeared incongruous, unkempt spies at Muybridge's Urania projections, sitting alongside eminent elderly anatomists, physicists and industrialists such as Helmholtz and Siemens. The brothers had been attracted to the projection event by the previous day's ecstatic newspaper accounts of Muybridge's first Urania appearance. Even more than Muybridge, the Skladanowskys' lives of the past decade or so had been spent perpetually on tour, across the cities of Northern Europe, as assistants to their father, who performed feats of magic, often using magic-lanterns to create eerie atmospheres of darkness and spectrality in order to enthral his spectators. The Skladanowsky family had migrated for survival to the ascendant urban furore of Berlin from the areas of acute poverty to the city's north-east that had never fully recovered from their depopulation in the brutal seventeenth-century Thirty Years War. Their father was now ailing and exhausted from his years of touring, and expected his sons to generate the family's income by initiating innovative touring spectacles of their own, in that era of Berlin's intense urban demand to sensorially witness the unprecedented.

The brothers were at the very beginnings of their career, and changed their names several times – as Muybridge did – appearing as the "Hamilton brothers" and with other names before eventually reverting to their original names. Over the four years following their attendance at Muybridge's Berlin event, they developed two different spectacles for the auditoria of Berlin and other cities: the staging of sea-battles involving collisions between miniature ships in tanks of water that evoked the locations of historical marine conflicts, and magic-lantern spectacles involving up to eight or nine simultaneous, overlayered slide-projections, with that proliferation paralleling Muybridge's use of multiple cameras for his sequential photographing of horses in motion in California in 1881. The brothers were able to work on their moving-image projection plans only intermittently over the years between 1891 and 1895, between their touring.

In the summer of 1895, after successfully assembling several film-cameras, the skilled but untrained Skladanowskys constructed the Bioskop projector in their workshop at a large warehouse in the Prenzlauerberg district. The projector used elements of Muybridge's Zoopraxiscope, such as a propelling circular handle, and otherwise shared the irreplicable idiosyncrasy of the Zoopraxiscope's conception. The Bioskop had two lenses for the projection of hand-cut 44.5mm celluloid strips, one lens projecting odd-numbered frames of a sequence, the other lens projecting that sequence's even-numbered frames; the manual revolution at speed of the frontal handle drove a sharply serrated metal disc in front of the lenses, alternately obscuring one and revealing the other, with the projector operating only barely at the speed at which the celluloid strips' images generated the appearance of movement for their spectators' eyes. The Bioskop (its name indicating that those spectators would "see life") was a double-projector in terms of its dual lenses, its dual celluloid strips, and also in the amalgam of potential malfunction and functionality which its lenses' oscillation generated. For their planned projections, the brothers shot brief sequences of human movement – dances and acrobatic spectacles – and staged trial projections for friends at a local venue, the Café Sello, before successfully convincing the proprietors of the Central Hotel's ballroom to hire them to project their work there, for the month of November 1895. From Berlin, they would then begin a moving-image projection tour that, over the course of fifteen months, traversed many of the cities of Central Europe in which Muybridge had projected his own work several years earlier, though the brothers' tour would extend also into Scandinavia and Holland. Like Muybridge, they became itinerant moving-image projectionists, their previous touring itinerancy undertaken for other spectacles now transformed into the work of film's projection.

As with Muybridge's Zoopraxiscope, the Skladanowskys' Bioskop projector survived, while many of their other materials were lost in the 1940s, burned during wartime bombing or dispersed in the invading USSR army's arrival in Berlin. For many years, the Bioskop was positioned, apparently abandoned, in a corner of the foyer of the Filmmuseum in Potsdam, close to Berlin, as though it were a detrital artefact, awaiting removal. While Muybridge's Zoopraxiscope possessed a horizontal orientation and was usually positioned on a table or flat surface for its projections, the Bioskop was vertically oriented. By 2020, when I made a further visit to the Filmmuseum, it had been placed within a large glass case in that foyer, with projectors manufactured in subsequent decades in Germany – the Imperator, the Matador-II – displayed on shelves above it. By contrast with those industrial, mass-produced projectors, the Bioskop appeared an outlandish, aberrant instrument of vision, with its jagged disc for the obscuring of one or other of its lenses. When I stood directly in front of its upright configuration with its two projection lenses at eye-level, it appeared to carry the aura of a human figure and face, as though that projector were the embodiment of an experimental filmic anatomy.

Some of the Skladanowskys' glass magic-lantern slides and celluloid strips also survived, along with the brothers' disordered scrapbooks and tour-schedules, conserved at the archive of Berlin's Pankow-district local history museum, close to the site of the brothers' workshop. There, the museum's curator eagerly encouraged me to take one after another of those slides and strips and to hold them in my hands, though the 130-year-old unique slides appeared extremely fragile, and would shatter if dropped. When I held up those slides to the light, I saw images of diabolical figures, houses on fire in mountainous landscapes, and catastrophes. Only a fragment of four celluloid frames from the Skladanowskys' very first film remained extant, with Emil filmed by his brother Max on a rooftop with brewery chimneys in the background, and while the upper two frames appeared pristine, the lower two frames had darkened, their heat-melted edges showing damage by soot or fire – most probably, the curator told me, from the determination of frenzied Soviet soldiers in April 1945 to burn or consume everything they could find at Max Skladanowsky's house, and at all other houses in their way, as they entered Berlin from its north-eastern perimeter (head-

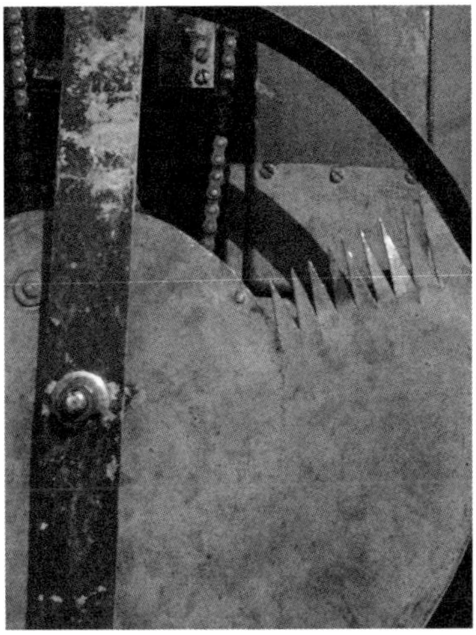

Ill. 16: The Skladanowsky brothers' "Bioskop" projector, Potsdam, 2020.

ing towards the central area inhabited by Muybridge's Urania auditorium). The sunlight outside the museum's window projected that fragment's traces of near-incineration into my eyes.

The entire programme of performing animals, magic and dance at the Central Hotel's "Wintergarten" ballroom on 1 November 1895 lasted for three hours, and the Skladanowskys' projections were positioned at the very end of it – when many of the 1,500 spectators who crammed the venue every night would already have left in exhaustion – and spatially at a peripheral area of the ballroom, away from the main stage. A scattering of round tables with chairs was arranged in front of the cloth screen. That projection had the aura of a tentative experiment, in contrast to the spectacle which the brothers had presented on the ballroom's main stage for the entirety of the previous month, October 1895: a sea-battle show, evoking the Roman-era battle of Alexandria (which preceded the deaths of Marc Antony and Cleopatra in 30BC), with roaring cacophonies, flares and fireworks, brutal collisions of battleships, and miniature models of human figures propelled from the impacted vessels into the water-filled tanks deployed across the stage, all of it vocally commentated from either side by the two brothers. That sea-battle spectacle had been a great success, while the brothers' November film-projection spectacle appeared more precarious, with its unprecedented technical demands and its unknown spectatorial response, though the ballroom's proprietors had paid them well for it, with a fee of 2,500 gold marks for twenty-three projections in all.

The spectators for the Skladanowskys' first public projection of celluloid film had no idea what they were about to witness, and no sense or guidance of how the projections would happen, especially since the brothers projected their films out of sight, with the Bioskop positioned directly behind the screen. That screen was saturated with water just before the projection began, in order to give maximal transparency and sharpness to the sequences. Although that "reverse" projection suggests an anomalous wrong-way-round approach, in the context of future histories of projection from behind the audience or from a projection-box, the brothers' decision appears to have been aimed primarily to mask the noise generated by the Bioskop's operation; a pianist played loud music in the auditorium itself to accompany the films.

As Muybridge had done with his Berlin projection events, the brothers showed several still slides at the outset, in order to enhance the exhilarating moment when the still image was abruptly transformed into the moving image. Then, for fifteen minutes, that audience witnessed the Skladanowskys' moving-image projections. Since the brothers' films were all extremely short – usually, four to ten seconds in duration – they were projected repeatedly, as loops, each for a minute or so. The corporeality envisioned by the audience within those films was an enthralling but also jarring one, with fragments of bodies propelled into relentless repetition, at a speed that barely reached the ocular threshold for the perception of movement. The bodies in those films – dancing, performing acrobatics, or wrestling – were simultaneously there, moving, while also oscillating into invisibility, into the danger of a stalled freezing, or into corporeality's death. The figures, often shot outdoors to maximise daylight, were accompanied by the prominent casting of their shadows. As with Muybridge's figures, those of the Skladanowskys' films often appeared to be pivoting and unbalanced, and held a distinctive, unsettling experience for their spectators. Their projection from filmic celluloid, as opposed to Muybridge's spinning glass discs (which were never intended to conjure any illusion of life in their figures' projections), imparted to them a unique spectrality which at least some of the spectators perceived as uncanny or even diabolical. Contrary to the accounts of the Lumière brothers' first public moving-image projections which took place in the following month in Paris, in the course of which spectators ran in alarm to avoid a filmic train arriving at La Ciotat railway station, the Skladanowskys' audiences were entranced by corporeality's gaps and voids in shattered time and space.

That sense of film's profound disquieting and darkening of perception appeared too in the numerous press-accounts of the Skladanowskys' projections. While uniformly positive in tone (at least in part the result of Berlin newspapers' reliance on advertising revenue from venues such as the Central Hotel, so that negative reviews almost never appeared in that era), those accounts intimated film's outlandishness: "The skilful technician [the projectionist: Max Skladanowsky] employs delightful moments of photography here, bringing them in enlarged form into representation, but in a living rather than stilted medium. How he does it, only the devil knows."[2] That account prefigures the determining unfathomability of celluloid film-projection through its future decades, as a mystery that entailed unforeseeable transmutations or malfunctions of perception. By contrast with its projections' volatile effects on human perception, the Skladanowskys' Bioskop itself invariably functioned smoothly, and the brothers would proudly assert in retrospect that they never had a breakdown (just as Muybridge's "operator", Ernest Webster, lauded the Zoopraxiscope's relentless dependability despite the rigours of Muybridge's extensive touring). The

Bioskop is still capable of projecting even now, over 120 years after the Skladanowskys ceased their projections.

At the end of their projections, the Skladanowskys – invisible up to that point – abruptly emerged into the auditorium. While their spectators witnessed the final film in the programme, *Apotheosis*, in which the brothers' filmed bodies appear from either end of the screen and approach its centre to receive the audience's adulation, the brothers themselves appeared from behind their screen, exactly replicating their bodies' gestures on that screen, with filmic corporeality and immediate corporeality pitched into raw juxtaposition. That adept manoeuvre enthused the startled but voluble spectators, who demanded (at least on the projections' first night) that it should be repeated several times. As with Muybridge's Urania projections of three years earlier, a rare dissenting voice was that of Anschütz, then gradually falling into financial failure after adapting his work from one-person viewing machines to larger-scale projections in auditoria (he had projected his glass-disc sequences of horses in motion at the Post-Gebäude hall in central Berlin in 1894, after returning from his final bout of rivalry with Muybridge at the Chicago Exposition in the previous year). Despite the evident disparity between his own projection media and subject matters, and those of the Skladanowskys, Anschütz made an attempt to take legal action against them after their 1895 projections, claiming they had stolen ideas from him (while he himself had stolen his ideas from Muybridge), but got nowhere with those two wily young magicians, who simply asserted that they had conjured filmic projection out of nowhere, in an unprecedented way, by themselves, in their Prenzlauerberg workshop, and had never heard of Anschütz and his sequences of horses; Muybridge, back in the UK by 1895, never showed any sign of awareness of the Skladanowskys' projections.

The Skladanowskys' success with their spectral, jarring projections at the Central Hotel's ballroom gave them the momentum to become itinerant moving-image projectionists, on tour across Germany and adjacent countries for fifteen months, with

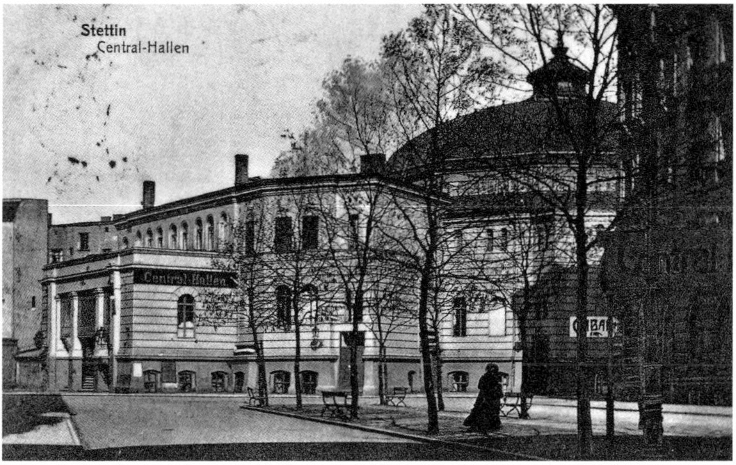

Ill. 17: The Central-Hallen venue in Stettin, c.1897.

intermittent spans back in Berlin, where they sought to maintain their tour's velocity by making new films, especially in the summer months of 1896, and by constructing a new projector (this time with a single lens) and new film-cameras. They also began to make brief, improvised films while on tour, as during their projections in Stockholm. Their moving-image tour, as with Muybridge's Central European tour of five years earlier, formed an irreplicable one, already vanishing as they pursued it. As film began to appear a viable industrial entertainment medium across 1896, other moving-image projectionists were now rapidly appearing in great numbers in the cities they passed through, and those projectionists' rapports with one another became adversarial rather than collaborative; planned projections by the Skladanowskys in January 1896 in Paris (at the Folies Bergère music-hall) were annulled through the intervention of their rivals, the Lumière brothers, along with a planned tour-appearance in London. The Skladanowskys now often divided their tour spectacles into two parts, with their celluloid moving-image projections as one part, and their sea-battle or other multi-dimensional cacophonic spectacles as the other part.

During the summer months of 1896 in Berlin, while their tour was halted, the Skladanowskys filmed the city. They needed new films, since the celluloid of their corporeal films of dance and acrobatics had worn out through repeated projections and those films had, in any case, now been comprehensively viewed by their audiences. The Skladanowskys were the first filmers of Berlin's urban space, and they shot the city multiply, with a relentless obsessionality which invokes Muybridge's proliferating experiments at his Philadelphia studio that generated the moving-image sequences for his Central European tour. The Skladanowskys filmed locations in the familiar area around their workshop in Prenzlauerberg, and also filmed in the adjacent Pankow district and made their own railway-station film in north-eastern Berlin, in the Schönholz district. Whenever human figures appear in those films, they are transiting the space at speed; the focus remains pre-eminently on city-filming. The Skladanowskys also travelled southwards into central Berlin as they had done in 1891 to attend one of Muybridge's Urania projection events, filming in the city's great squares and avenues such as the Alexanderplatz, Unter den Linden and Friedrichstrasse (where the Central Hotel was located). They envisaged a new projection spectacle entirely composed of those city-films, and proposed it for a repeat appearance at the Central Hotel's ballroom at the beginning of 1897, but by that time, Berlin was engulfed with its aspirant projectionists – often serving simultaneously as the agents of filmmakers based in other cities, such as Paris, London or New York – who operated devices with far greater technological complexity than the Skladanowskys' hand-built projectors; the Skladanowskys' proposal for a second month-long series of Central Hotel projections was dismissed. The industrial standardisation and normalisation of film projection was already beginning, and the outlandish idiosyncrasy of the Skladanowskys' projections, with their aura of insurgent devilry, now appeared obsolete.

The Skladanowskys had assembled the first programme of city-films, from their 1896 filming of Berlin, but had no opportunity to project that programme in Berlin itself. It also proved difficult for them to generate another tour of their projections, with the result that their attempts to schedule that new tour became reduced to a single engagement in one city. In March 1897 (two months before Muybridge travelled to St Ives for his own final projection event), the brothers travelled several hours north-east by train to Stettin, Berlin's Baltic port-city, to project their Berlin city-films for two weeks as part of a programme alongside other spectacles. At that time, Stettin was a rough marine city of alcohol and uproar, its frenzied animation collecting around the quays of the river Oder. It was the end of the line, and a last-ditch urban tour-stop which most itinerant performers preferred to avoid; film remained almost unknown there. The Central-Hallen venue, reached via a short tree-lined esplanade from one of the city's avenues and then by climbing its colonnaded entrance's marble stairs, held an elegant, domed and semi-circular auditorium that resembled Muybridge's Berlin projection venue, with its gilt ceiling and lavish chandeliers, though it was far larger than the Urania lecture-hall, seating 2,000 spectators on its many balconied levels. The venue's entrance was surrounded by exclamatory announcements of forthcoming attractions; a new programme always began at the Central-Hallen's auditorium on the 1st and 16th of each month. The venue had been constructed, eight years earlier, primarily for variety spectacles, its auditorium designed with multiple audience perspectives, and many of the seats were positioned in such a way that a clear view of a fixed projection screen would have been impossible.

Once they had arrived in Stettin, the Skladanowskys made a film, as they had in other cities on their previous year's tour. But rather than a corporeal sequence showing a sudden fight or confrontation between a group of figures, as with their previous films, they instead made a Stettin city-film to accompany their programme of Berlin city-films, crossing to the far side of the river Oder to film the main quayside's intense uproar, with many steam-propelled ships churning the river's water as they docked or departed, alongside the teeming figures of sailors, prostitutes and hawkers amassed on that quayside, and with a cityscape of medieval buildings and towers, industrial chimneys and church steeples extending across the rising hillside behind the quay. Stettin's urban space had never previously been filmed. The brothers immediately developed their film (thereby anticipating the agile in-transit development and projection of films that is associated with later filmmakers such as Vertov and Medvedkin who were active in the USSR's first decade) and had it ready to project to their audience at the end of their first night's programme at the Central-Hallen venue. The Skladanowskys were now projecting a greater width of celluloid, 63mm, from their single-lens "Bioskop-2" projector, which gave their films a marginally greater speed and resolution than their previous projections. Press-accounts of their first Central-Hallen projections on 16 March 1897 emphasise that it was the Stettin film, rather than the relatively alien Berlin-films (which many of the disorientated spectators could not locate or recognise, either through film's determining strangeness or because they had never travelled there), which especially enthused the audience. But, as with the Skladanowskys' Central Hotel projections, those accounts also highlight the projections' uncanny spectrality, with film's rendering of Berlin into a near-depopulated city intermittently traversed by frantic ghosts.

The Skladanowskys' one-stop tour to Stettin had been a last-chance, last-gasp filmic journey, and after the final projection of their city-films programme there on 31 March 1897, they never projected again. Their tour's cancellations and the expenses of constructing new devices had depleted them financially. The Berlin municipal bureaucracy arbitrarily cancelled their official license for public projections, on the grounds that there were now too many film-projectionists operating in the city, though the Skladanowskys had been the very first of those projectionists. They were supplanted by film's ascendant industrialists; it would take a further ten years for those industrialists to develop auditoria specifically for moving-image projection – as Muybridge had already done in 1893 with his Zoopraxographical Hall in Chicago – but by the end of the twentieth-century's first decade, Berlin would hold hundreds of cinemas, many of them immense and palatial, each equipped with their professional film-projectionists. Following their tour-exhausted father's death, soon after their return from Stettin, the Skladanowskys fell out in an inheritance quarrel of such torn-apart acrimony (as in a sequence from *The Brothers Karamazov*) that the brothers never spoke to one another again. They abandoned film projection; Emil Skladanowsky, whose work had always been primarily that of arranging the brothers' tours, became the manager of a water-spectacle touring company, and Max Skladanowsky, who had always operated the Bioskop, turned to the production of photographic flick-books.

The Skladanowskys had already experienced nearly four decades of oblivion when their work was unexpectedly resuscitated in 1933, shortly after Hitler's accession to power, by Goebbels, Hitler's Minister of "public illumination and propaganda"; Goebbels immediately welded his own passion for film to his nationalistic imperatives, presenting Max Skladanowsky as Germany's pre-eminent filmic innovator and organising extensive projection tours during which Max Skladanowsky would demonstrate to audiences the operation of his Bioskop, before the brothers' films were shown, now transferred to standardised celluloid formats and projected by professional projectionists. At a gala presentation in May 1933 at the 2,000-spectator capacity Beba-Palast Atrium cinema (whose immense, domed auditorium in Berlin's Wilmersdorf district resembled that of the Stettin Central-Hallen venue, and was also destroyed by aerial bombing a decade later), Goebbels stood directly beside the projection-screen to give a live vocal commentary on the Skladanowskys' films. Goebbels diabolically recast the Skladanowskys as glorious nationalistic figures, rather than as the young, audacious devils who, largely via their nonchalant thievery of Muybridge's work, had once magically initiated celluloid film-projection. Hitler himself viewed the Skladanowskys' films at a private projection in 1935, though his preference was for extravagant musical films and it appears unlikely that he would have been enthralled by the Skladanowskys' extremely brief, experimental films of corporeality and urban space.

It proved almost impossible to view the Skladanowskys' film of Stettin: their final film, and the last film they projected, on 31 March 1897 at the Central-Hallen auditorium. The city of Stettin vanished in 1945, largely destroyed by the previous years' intensive aerial bombing – including the detonation of the Central-Hallen venue itself, which received a direct hit, as though the UK Royal Air Force bomber-crew had aimed specially at it – and then overlayered by the Polish city of Szczecin, at Stalin's obstinate caprice, since it was located on the west bank of the river Oder, which had been intended to form north-eastern Germany's new boundary with Poland. As with many of Stalin's urban upendings, one population was expelled, and another assigned for the city's new inhabitation. Szczecin's current cultural administrators and archivists appeared to have limited interest in an 1897 projection in a disappeared city, nor in the first momentary filming of that annulled city's urban space. I finally ascertained that the German state film-archive had a preserved celluloid copy of the Skladanowskys' film, though it was an old and fragile print, potentially the last one in existence, and most likely dating from the 1930s resurgence of interest in the Skladanowskys (though the Stettin film was not part of Max Skladanowsky's Goebbels-sponsored Nazi-era tour, and may have been copied from the original film either accidentally, or solely for the sake of obsessive completion); it could only be projected for me in Berlin at the cinema auditorium of the German Historical Museum, with its optimum projection equipment that would cause the film no damage, and even then, only once.

Passing through the museum's foyer (where an aspirant suicide bomber, Rudolf Christoph Freiherr von Gersdorff, had once attempted to assassinate Hitler with explosives hidden in his pockets), after its closure for the day, I entered its Zeughaus cinema's auditorium, and met its elderly projectionist. He had been employed there since the era when the building had been part of East Berlin, and I remembered that I had met him once before, in 1993, when he had projected a complete season of Murnau's films, and had often had to enter the auditorium to explain to the spectators that a dilapidated, unique film-print had split-apart and now needed careful suturing. He took me into the projection-box and with evident pride showed me his line of projectors, able to project film-reels of any dimension, then prised open the small film-reel's metal can. He had already cleaned the film and ascertained that it was still possible to project it. Inspecting it again closely, with reverential care, he told me it was unlikely to have ever been projected in the last eighty or more years. He was uncertain if this was the complete film shot by the Skladanowskys, or only a fragment of it, but it would certainly take between three and four seconds to project, and he was authorised to project it only once, so I needed to be ready.

I sat in the auditorium and the film began, appearing as though in a sudden urban burst. I saw the city of Stettin, filmed from the eastern side of the Oder river, its water agitated by the ships' propulsions. I had an instant to seize the human figures' dense movements on the quayside behind the ships, and to pivot my eyes upwards to the dense accumulation of Stettin's buildings and towers, then the city abruptly vanished, the screen emptied, and I heard the projector shut down behind me.

From Berlin, I took the single-carriage train on its buckled single line, north-eastwards through the depopulated lands that had over centuries been the site of multiple wartimes' razings, to Szczecin. The Skladanowskys must have reached and left the city by that same route, along with their Bioskop-2 projector, projection-screen, film-cameras and film-reels, in the last moment of their existence as transient moving-image projectionists. The railway station was located directly alongside the river Oder, its former quaysides now almost empty, with no ships visible. I crossed the bridge to the river's far side, ascertaining the exact location from which the Skladanowskys' Stettin film had been shot. The quayside area on which their camera had focused had been almost entirely obliterated a half-century later, and remained constellated with voids between concrete towers and shoddy reconstructions of merchants' houses; one medieval circular brick tower that appears in the Skladanowskys' film – the "tower of the seven cloaks" – had survived, directly alongside a multi-levelled concrete highway-bridge. The only gestural action on that quayside was that of a restless teenaged gang, vomiting and fighting. The city I had seen in the Skladanowskys' Stettin film had been erased. Even so, everything that had been filmed in 1897 remained simultaneously there in the contemporary instant, in ineradicable filmic intimacy with urban space, in its moving-image survival and its potential revivification via projection, between film and the eye.

I crossed the river and took the Nowa avenue towards the unmarked site of the Central-Hallen auditorium. On the way, I passed the Pionier cinema which announced itself as the oldest continually operating cinema in the world, having begun in 1907 as the "Helios-kino" (using the same name, that of a god of the sun and its illumination or darkness, which Muybridge had assigned to himself for his San Francisco street-photography at the end of the 1860s, when he had itinerantly traversed that city with his cameras, immediately developing his work in a mobile cart which he called his "flying studio"), passing from country to country, from Germany to Poland, and from celluloid-film projection to digital projection. That cinema had come into existence a decade after the Skladanowskys had filmed Stettin and projected their work at the Central-Hallen's auditorium. Further north, within a district of 1880s tenements which resembled those surrounding the Skladanowskys' Prenzlauerberg workshop, I turned off the avenue to the site of the esplanade that had been the approach to the Central-Hallen venue. Only one side of the line of once-majestic five-storey buildings that had boundaried that esplanade now remained, their facades now eroded and striated, with the other side replaced by a wing of a derelict school, whose main buildings now also blocked the esplanade's course. An un-tarmaced path marked with signs of prohibition against entry passed to one side of the school and the empty yard behind it, and beyond that yard, I scaled a high rusted gate and found the wasteland of the Central-Hallen venue's site.

All of the Skladanowskys' projection sites, in Berlin and Stettin, were destroyed: Berlin's Central Hotel, also by aerial bombing, in 1944, and the Café Sello, where the brothers made their experimental projection tests, demolished in 1927, that site later holding a cinema which was itself demolished and turned into a supermarket. I entered the black-earth terrain of the Central-Hallen site. A few abandoned cars had been left at its northern edge, long ago, but otherwise, it appeared entirely deserted, not even populated by Szczecin's raucous teenaged street-gangs. At the wasteland's entrance, the base of the venue's marble staircase was still deeply embedded into the earth, seemingly impossible to dislodge, having demonstrated its resilience to the direct-hit explosion which had otherwise razed the building on the night of 20 April 1943. The Central-Hallen venue's mass of debris had been carefully assembled and arranged in a great heaped line, but then never cleared; that line extended from the heart of the wasteland and ended alongside the fire-blackened rear-walls of tenement buildings whose blown-out windows remained voided.

I approached the bank of debris, up to seven feet or so in height, and climbed onto it to look directly down into the deep fragmentary history of the Skladanowskys' final projection-site: I could see fire-blasted shards of ornate green-glazed tiles from the Central-Hallen's entrance facade, some of them lettered with elements of its name and with announcements for forthcoming attractions, shattered glass from the venue's windows and dome, ventilation filters, chimney turrets. I picked up a handful of that serrated debris, in its amalgam of brick, bronze and glass, and sieved it through my fingers. It seemed that a strange and intractable interdiction had been cast, forbidding

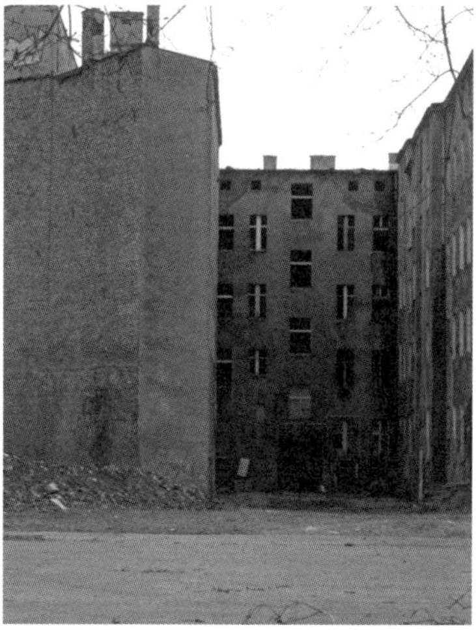

Ill. 18: Debris of the Central-Hallen venue, Szczecin, 2020.

the city's population ever to clear or even enter that cinematic terrain, as though it needed to be maintained intact as a monument to moving-image projection's devilry, innovation and aberrance, embodied in the figures of the Skladanowsky brothers, two young rogues who arrived in the city in 1897, filmed its quayside, projected their filmic sequences in its grand auditorium, then left the city behind, while cursorily renouncing their own status as projectionists at that same moment. That site held the raw detrital traces of the brothers' adept capacity both for experimental invention and for self-willed catastrophe, as initiating figures of moving-image projection, all the more vital for their projection sites' losses and for their work's near-oblivion. I let the last debris of the Central-Hallen building's destruction slip through my fingers. Then I climbed back over the rusted gate and left the wasteland of the Skladanowskys' projections behind.

Notes

1. This chapter was primarily researched in the archives and collections in the Berlin region which hold the Skladanowsky brothers' materials: especially, the Potsdam Filmmuseum, which has the "Bioskop" projector, the Deutsches Technikmuseum, which has the Skladanowskys' film-cameras and the "Bioskop-2" projector they used in Stettin in 1897, and the Pankow local-history museum, which has an uncatalogued extensive collection of the Skladanowskys' celluloid film-strips, glass-discs, scrapbooks, schedules and other artefacts. I'm grateful to everyone, especially Sara Piazza, who helped me access those materials, many of which are unknown.

2. Anonymous account in the *Staatsbürger Zeitung*, Berlin, 5 November 1895, page 5.

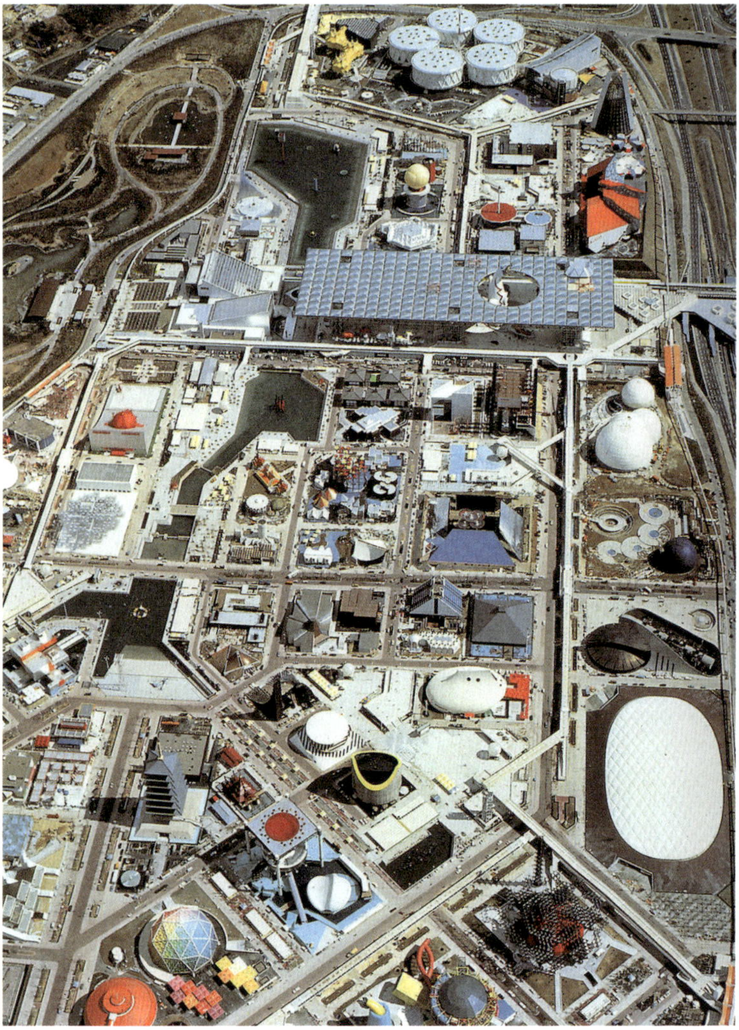

Ill. 19: The Osaka Exposition site photographed from above, 1970.

Projections of the Osaka World Exposition, 1970

The Osaka World Exposition of 1970, held seventy-seven years after Muybridge had operated his Zoopraxographical Hall at the Chicago World Exposition, was focused almost entirely on moving-image projection experiments and on the construction of architecturally innovative pavilions as auditoria for immersive projection events – demanding the innovation of new, large-format film-projectors – whose spectators were intended to experience unprecedented corporeal and sensorial upheavals, especially in their relationship to the projection screen, towards which they were propelled at speed on hydraulic seating or towards which they rotated their bodies and eyes while listening to an amplified vocal commentary. 64,000,000 spectators – half of the population of Japan, alongside immense numbers of international visitors who travelled to witness the event – attended the Osaka World Exposition over its six-month span: the greatest number of people that had ever participated in one event. Each of the many moving-image projection auditoria that constituted the Exposition grounds was visited by around 12,000,000 spectators. The directors of the Exposition conceived of its site as an "experimental metropolis" able to allow its inhabitants to reflect on the tumultuous decade which had just elapsed, as well as to focus on the future of corporeality, urban space and moving-image media as the pivotal elements of human life.[1]

In many ways, the Osaka Exposition constitutes an epochal "projection event" in a parallel way to that in which scientists conceive of "extinction events": seminal and formative occurrences or shifts, with catastrophic dimensions, that mark engulfing transformations and also determine the future of human (alongside other forms of) existence. An extinction event is fundamentally non-linear and often leaves no clear evidence by which it can be reconstructed or even understood. It may even be subject to widescale oblivion. As the greatest-ever gathering of the human population, explicitly to experience the sensorial and ocular impacts of Muybridge-derived moving-image projection acts, the Osaka Exposition holds a parallel advanced resistance to its linear reconstruction or understanding, or to its cogent archiving. That Exposition projection event, in its corporeal excess, forms a fundamental mishap and mutation in the ocular and movement-focused development of human vision, as well as a vast-scale experimental urban project demanding immense financial resources for its anti-ecological conception and realization. It also possesses an extreme contradiction in its projections' undertaking by experimental artists, architects and technologists, in tension with its status as a closely controlled urban entity designed for maximal state-directed surveillance and for the promotion of corporate power.

The Osaka Exposition drew on previous World Expositions' focus on experimentation with moving-image projection, such as that widely undertaken at the preceding Montreal Exposition of 1967, but its profound relationship is with Muybridge's innovation of the Zoopraxographical Hall as the conceptual breakthrough that allowed moving-image projection to inhabit its own uniquely designed space, as an exploratory project intended to unleash an unprecedented spectatorial experience. Muybridge's inspiration is often in directly inverse proportion to his work's profile and exposure; it thrives on its own shadowed oblivion. Muybridge's catastrophe in Chicago (in which a rogue from Naples exploited his financial vulnerability to seize his "concession", in order to supplant and prematurely end it, while probably fearing that the black-suited, bearded killer, Muybridge, with his reputation for a volatile fuse and a short temper, would again reach adroitly for his gun) is fundamental to his work's subsequent pervasive instigation of moving-image inhabited urban environments and to his pivotal inspiration for contemporary digital moving-image art. Muybridge's work and its inspiration are always malfunctional, unbalanced and vertiginous (as with many of the figures in his moving-image sequences), existing beyond and against linear time, operating in the zone of simultaneity and immediate gestural action that his multiple-camera sequences envision.

At the moment when he died in his Kingston-on-Thames back-garden in 1904, Muybridge was largely forgotten and no press media whatsoever recorded his death; only a decade or so earlier, he had been the focus of saturated media attention for his maximally acclaimed work as a transient moving-image projectionist in Berlin, Munich and Vienna. And only a decade or so after Muybridge's death, as his architect Wing's obituary in the *Chicago Tribune* indicates, he had already been widely elevated to the status of the originator or inventor of film (a status Muybridge never claimed, determined instead at the end of his life, through his decade of work on his scrapbook, to reconfigure and streamline his life's work as art that had experimented with moving-image and photographic technology). His work, across the twentieth century's later decades, deeply inspired both individual artists, such as Francis Bacon and Sol LeWitt, and large-scale public audience-focused projection events, pre-eminently in the form of the Osaka Exposition. The negligible and the maximal – and their abrupt oscillation – are dynamically conjoined dimensions of Muybridge's work.

The immense scale of his influence on that Exposition's conception – in the expansion and transferral of his auditorium-located moving-image projection innovations of 1893, into the projection pavilions of that experimental megalopolis in Japan in 1970 – also resonates with the acute scarcity of traces from Muybridge's own projections (or vocally fabricated, non-existent projections) in Japan, during the largely unknown interval between his expulsion from his Chicago auditorium and his return to England. The great power of film, digital media and moving-image culture is derived directly from its basis in untruthful concoction and volatile, non-linear oblivion. The eminence of both Muybridge's work and the Osaka Exposition as inspirational presences for contemporary art are unrelated to public consciousness or awareness, and exist contrarily in an inverse relationship to that level of consciousness. The more that a body of work or an epochal event is consigned to oblivion, the more determining and inspirational it becomes, rendered inassimilable and invisible to its dissolution or co-opting in media culture, and located beyond the linear strata of cultural influence. Whenever public attention focuses on those formative projects in moving-image projection histories, as for example during the Tate Britain art-museum's Muybridge exhibition in 2010-11 (which had travelled from the USA, from the now-dissolved Corcoran Gallery of Art in Washington DC), and around the announcement in 2019 that Osaka would host a successor event to its 1970 World Exposition, in 2025, those entities' integral mysteriousness and lostness is even enhanced and consolidated. In the interconnected endurance and future existence of Muybridge's moving-image projection experimentation in Chicago in 1893 and of the 1970 Osaka Exposition event, time itself lapses, or disintegrates, then resurges in an opposed direction.

Japan had undergone engulfing upheavals throughout the 1960s, across corporeal and urban dimensions, and those transformations rendered it readily exposed to the future-oriented sensory experimentation which the Osaka Exposition's moving-image projection auditoria envisaged. Urban time itself was overloaded and contaminated, in the recent-past incineration of Japan's cities of 1945, through their subsequent widescale chemical-industrial and traffic-generated pollution in their 1950s reconstruction, and via their 1960s brutal street-battles of student protesters against riot-police armies, focused in Osaka as well as Tokyo, in defiance of the USA's resilient military, cultural and political power following its 1945-52 Occupation of Japan. Corporeal time too was one of great shifts and instability in the two decades leading up to the Osaka Exposition, with the mutations of Japan's human bodies in new dietary regimes, through intensive working and entertainment consumption practices, and via exposure to air toxicities. The relentless televised media culture of the 1960s and that decade's incessant social protests and affrontments also left Japan's post-warfare bodies and cities already exhausted and disorientated, and led finally to a unique puncture point in time which the Osaka Exposition then occupied and inhabited, with its proposal to its immense audiences of a conjuration of renewed bodies and unprecedented urban environments, transmitted directly by its pavilions' moving-image projection experiments. In that sense, 1970 Osaka resonates intimately with 1893 Chicago in that earlier project of generating an experimental urban terrain at the city's edge (Chicago's southern edge, adjoining Lake Michigan, while the Osaka Exposition was created through the deforestation and levelling of the city's northern periphery), designed as an outlet of instructional culture and exotic entertainment for that over-concentrated tenement city of slaughter and commerce.

The Chicago Exposition and the Osaka Exposition were also conjoined in their inhabitation by pervasive security forces, from the 1893 "semi-military" police which patrolled the Jackson Park main site but steered clear of Muybridge's Midway Plaisance annex, to 1970's many thousands of guards, often recruited from local gangster clans and equipped with prototype technologies of mobile telephones and surveillance cameras, who kept watch over the seething crowds and also reinforced the division between unruly urban Osaka and the Exposition site itself. The Chicago Exposition appears far more directly linked to the adjacent city and its population, evidenced in photographic documentation of vast, ecstatic crowds in the city-centre celebrating the event's opening and through that Exposition's use of central-city venues beyond its own parameters, as with Muybridge's ill-attended Exposition lecture-projection of July 1893 at the city-centre Art Institute of Chicago, following his expulsion from his own auditorium. By contrast, the Osaka Exposition's visitors from the adjacent city often felt a reluctant obligation (a compulsory obligation, for its children) to attend that new manifestation of future-Japan; after their day of being captivated by successive exposures to the event's many

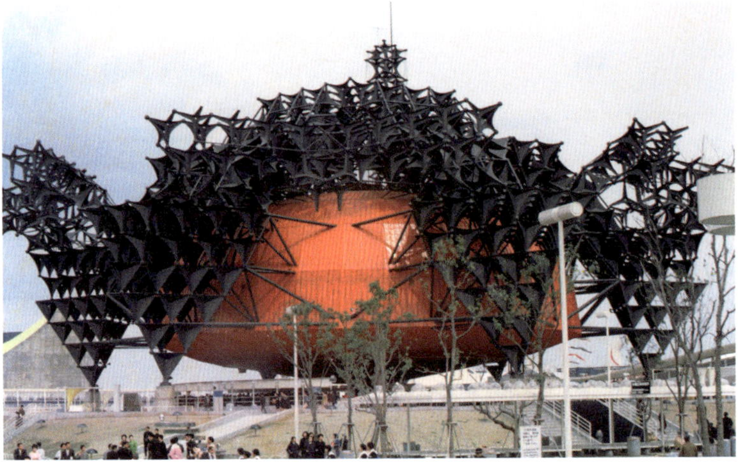

Ill. 20: The Toshiba IHI pavilion at the Osaka Exposition, 1970.

sensorially exacting moving-image projections, those visitors often responded in exit questionnaires that, on their release from the Exposition grounds, they now felt an aroused “antipathy” towards the exhausting experimental event which they had just experienced.[2]

The Osaka Exposition’s directors and corporations decided on the construction of an experimental city of pavilions to be inhabited by immersive moving-image projection environments, as the event’s pre-eminent content and preoccupation, as though in a near-collective hallucination, guided in beguilement by the detrital spectre of Muybridge’s Zoopraxographical Hall. Several of the corporations made documentary films about the process of commissioning and constructing their pavilions, and the documentary crews meticulously filmed horizontal tracking shots of the corporations’ all-male, all-elderly executives as they voted in their boardrooms on whether or not to commission experimental moving-image projection environments for their pavilions, with most executives raising their hands in assent, but with one or two obstinately refusing, as though those executives were re-enacting one of Muybridge’s sequences of bodies in movement in which he intentionally incorporated occasional misfired images of malfunction. Although several Exposition pavilions were built in a classical oblong form whose pillared facades resonated with that of Muybridge's own venue, most pavilions took the form of immense, plastic-coated domes, inspired in part by the geodesic domes of Buckminster Fuller (who had participated in the 1967 Montreal Exposition) but more immediately by the innovative young Italian architect Dante Bini, who specialised in rapidly air-inflated domes and visited the Exposition grounds to supervise the construction of a series of subsidiary domes arranged around the corporate Fuji Group's main projection pavilion. Bini (a close friend of Michelangelo Antonioni) knew moving-image history and Muybridge's work well, and when I visited him in Tuscany in 2016, he told me with delight: "The entire Exposition maybe was Muybridge's revenge for Chicago." Bini successfully sublicensed his domes' design to Japanese corporations for multiple usages, including moving-image projection, and they proliferated throughout Japan's cities in the early 1970s.[3]

Bini's commission resulted from the collective determination of the Exposition's directors and corporations to commission what would be viewed as the most unprecedented, startling contemporary architecture for their pavilions, and then to install artists' moving-image projection experiments within those pavilions, usually allowing the artists total freedom to film and project whatever they wanted after giving them a vague future-oriented "theme" to work with. The corporations' sole intention was to amass prestige for their companies from the aura of sensational innovation generated by their pavilions. Although the Exposition also held many "national" pavilions with instructive and educational remits, its principal emphasis (and the primary focus of its visitors' attention) was on the moving-image projection pavilions commissioned by Japan's powerful corporations, such as Toshiba and Fuji Group, which offered new and immersive sensorial and corporeal experiences (awe, ecstasy, disorientation, vertigo, exhilaration) that, in many ways, expunged and reconfigured Japan's exhausted, wartorn bodies of the previous decades.

While the Chicago Exposition's main site had been composed entirely of instructional pavilions representing the USA or its states, or other countries (including Japan, with its facsimile temple-pavilion), the Osaka Exposition contrarily allowed the infiltration into its main site of pavilions commissioned and sponsored by Japan's corporations, so that its spectators' transits through or around that experimental city (on foot or via monorail) generated an acute sensory oscillation between the national pavilions' educational focus and the corporate pavilions' sensorially engulfing moving-image projection environments. The national and corporate pavilions' locations were intimately mixed, and that main site's locus was formed by a large plaza designed by the architect Arata Isozaki, whose work's reputation held an aura of electrifying innovation at that time, with his designs for cities emerging from Tokyo's bay and of cities embodying ruination; Isozaki had developed many collaborative involvements with Japan's experimental artists, enabling their subsequent co-opting into the Exposition's corporate project. That plaza held an immense, human-faced tower, the "Tower of the Sun", conceived by the artist Tarō Okamoto; its prominence attracted the attention of Japan's emergent terrorist factions, and one terrorist successfully eluded the site's many security-guards to scale the tower, then emerged through its face's eyes, and repeatedly yelled "Down with Expo" to the visitors below, as though engaged in an experimental corporeal projection that necessitated a sustained vocal commentary.

The Osaka Exposition's main site had an annex on its eastern side, "Expo Land", which explicitly evoked the Chicago Exposition's Midway Plaisance annex. All of the Expo Land concessions were rented and operated by non-Japanese corporations and entrepreneurs, just as the Midway Plaisance attractions were contracted to non-American (mainly Italian, German, and English, in Muybridge's case) operators. As on the Midway Plaisance, the pivotal space of Expo Land was occupied by a Ferris wheel, which drew the greatest crowds. The Pepsi-Cola Corporation commissioned its domed Expo Land-located "Pepsi-pavilion" with an immersive metal-foil interior whose reflections inverted and distorted the movements of its visitors' bodies, along with moving-image projections and experiments in vocal cacophony. A Czechoslovakian-operated moving-image pavilion generated an explicit evocation of the 1890s era of Muybridge's Chicago Exposition projections, with human figures abruptly emerging from its screen in mid-projection, as with the Skladanowsky brothers' corporeal emergence from either side of their moving-image screen at their 1895 Berlin projection events. Paralleling Muybridge's Zoopraxographical Hall, that largely unvisited Czechoslovakian attraction closed early in the Exposition's span, and the Expo Land annex never held the dense crowds which the Midway Plaisance was able to, once all of its attractions became fully operational; visitors complained that Expo Land was remote and difficult to access, and that they had already undergone intensive moving-image projection experiences within the main site's pavilions, since all of the Exposition's projections were extremely similar and operated on the principle of incessant repetition which was also pivotal to Muybridge's work.

The Osaka Exposition's corporations attracted the involvement of Japan's experimental artists, filmmakers and architects through the lavish, unlimited budgets and fees which they were able to offer. Even artists previously preoccupied with social insurgency, revolution and uncompromised resistance to corporate culture readily agreed to participate. The Exposition's corporate commissions presciently invoke contemporary dilemmas about the corporate sponsoring of experimental moving-image (and other) artists' works, notably because many of the commissioning corporations' directors were explicit about their aim to generate an amalgam of corporate culture and art that would annul and negate the kinds of insurgency which Japan had experienced over the previous decade. Those devil's alliances operated in all of the Exposition's corporate pavilions, but were all the more rawly evident on the Expo Land site. Many prominent young innovators in moving-image projection, such as Toshio Matsumoto, had their future work creatively wrung dry or ended by their participation in the Osaka Exposition. Muybridge, too, had little future for his work after his involvement in the Chicago Exposition, but, by contrast, he had operated his Zoopraxographical Hall in a far more solitary, vulnerable way; uncommissioned (though he declared he had been invited to participate), and paying all of his costs himself, his unprecedented experimentation in fixed-space moving-image projection had manifested itself in fierce, uncompromising autonomy.

Spectators at the Osaka Exposition's experimental metropolis queued for hours to enter the domed moving-image pavilions and experience engulfing sensorial and ocular disorientation for durations of around fifteen to twenty minutes; the film documentaries of the pavilions' corporate owners, such as Toshiba and Midori-kai, show those patient lines under awnings protecting them from sun or rain, and gazing at starting-time placards whose form derived from those which Muybridge installed outside his Zoopraxographical Hall. The sheer numbers of spectators entering each of those pavilions – tens of thousands each day, millions over the Exposition's span – recall Muybridge's audiences as envisaged in Wing's preliminary architectural drawing and in Muybridge's decision to incorporate "overspill" ticket-booths and entrances into his building's sides. Muybridge must have been remembering his excessive Central European tour audiences who had crammed venues such as the Munich Odeon and whose beyond-capacity corporeal excess and adulation directed towards Muybridge were documented in his events' press accounts. In the films of those Osaka Exposition pavilions, nobody is loitering or hesitating about entry, nor strolling past to negate spectatorship, as in Hartmann's photograph capturing the Zoopraxographical Hall; spectators outside the corporate moving-image pavilions appear utterly determined to enter, for their abrupt immersal

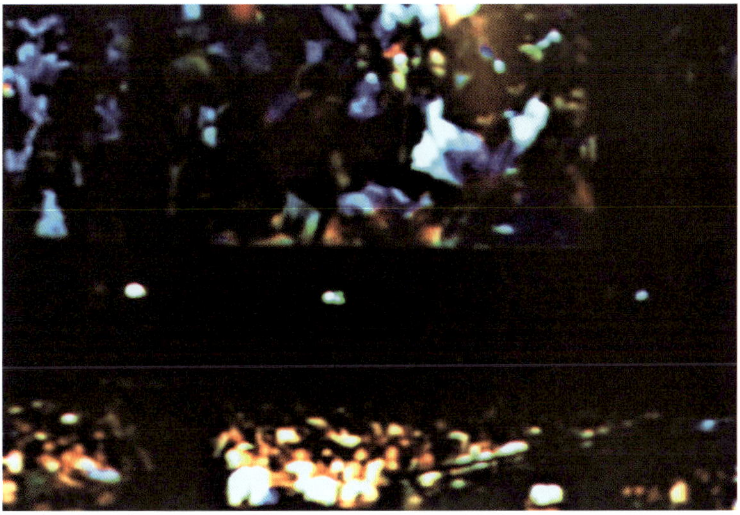

Ill. 21: Spectators propelled at speed on "rotary platforms" below the Toshiba IHI pavilion's immersive projection screens, Osaka Exposition, 1970.

in corporeal and urban moving-image sequences. At the projections' endings, the spectators were rapidly jettisoned through exit-doors by uniformed ushers, in an extreme contrast with their decelerated entrances into the auditoria.

Once inside, the spectators had entered the domain of the pavilions' projectionists: technically highly-skilled figures who had often also shot the sequences which they were now about to project, in exorbitantly large celluloid gauges and through the deploying of multiple experimental projectors positioned at intervals, each in its own projection-box, around the auditorias' concave interiors. As an exception to the habitual solitude of most projectionists, those figures operated in collaborative teams, communicating silently with one another via adept gestures or via prototype mobile phones between projection-boxes. The teams of projectionists were primed to begin projecting in synchronicity at the exact instant when the packed audience of a thousand or so spectators had all entered the auditorium, usually at a hurried run. The large-scale celluloid format of the films to be projected (up to 210mm, at the Fuji Group pavilion) made it necessary for the reels to be led horizontally into the projectors' gates, to achieve a stable projection. As well as the members of the projectionist teams who operated or oversaw the projectors, others sat intently in front of multi-screen monitors which enabled them both to ascertain the projection's uniform sharpness across the immense-format screen in the auditorium, and also to assess its sensorial impacts upon its crushed-together, awe-struck spectators, who frequently lost consciousness in their ecstasy or exhaustion and had to be urgently evacuated by the ushers.

Inside the auditoria, the spectators either stood, or sat in stasis, or were propelled around the pavilions' inner space on mobile, whirling platforms. Those who stood, gripping metal bars to avoid falling, were compelled to revolve their bodies and heads in wildly erratic and rapid movements as they followed the figures being projected, also in violent movement, on the immersive screens. Cities of concretised overpasses and immense towers flew by on those centrifugal screens. The corporate film-documentaries shot in densely packed auditoria such as the Midori pavilion show the standing audiences' mesmerised, bewildered figures and faces in movement. Seated spectators, filmed in the Pepsi pavilion, appear to be intently analysing the movements of the projected figures, as Muybridge must have intended his own (probably seated) audience at the Zoopraxographical Hall to do; between projected sequences, those Pepsi pavilion audiences scrutinise their own and others' elongated and upended corporeal figures, in their distortion by the pavilion's foil-mirrored interior, while cacophonies erupt. At the Fuji Group pavilion and Toshiba IHI pavilion, the spectators flew through the air as their seating lifted vertically from the ground and then traversed the space of projection, propelled by elongated, hydraulic devices. At the Fuji pavilion, seats were secured onto "revolving platforms" for their strapped-in spectators' vertiginous transits, while at the Toshiba pavilion, segments of the spectators' seating were attached to multiple "rotary platforms" which then veered rapidly around the space, sometimes impelling the spectators directly towards the illuminated screen, as though they were about to enter it, pulling back only when the spectators' hands could almost touch that screen and their eyes were in extreme proximity to it. During their platforms' revolutions, the spectators appearing in those documentary films burst into screams, or maintain an open-mouthed stunned silence, though in the Toshiba pavilion's printed programme accompanying the projection event, its directors appear contrarily to have envisaged an advanced act of communication taking place between spectators during their experience: "The idea of the Global Vision Theater was to create a place where spectators could communicate with people of other countries through images and think mutually about the happiness of mankind in a future society with 'hope'."[4]

The commissioned artists who made the films for the pavilions' projection spectacles undertook their work with an evident oblivious disregard for the guiding "themes" given to them by their corporate masters, often in such cryptic forms (as with that of the Toshiba pavilion: "Light for Man") that they instead seized their insurgent and lavishly-paid opportunity to film sequences which invoked the aura of urban and corporeal contamination, and of riotous confrontation and emergent terrorism, that had been unleashed globally and especially in Japan at the 1960s' end. The projected sequences showed apocalyptic wasteland devastations, dystopic cityscapes, grotesque devils and simulated orgies. Above all, they showed conflagration, in multi-screen sequences of children or youths whose faces are abruptly shadowed and replaced by images of fire, so that when the spectators were propelled towards the projection screen on their "rotary platforms" at the Toshiba pavilion, it was as though they were being drowned in fire. The pavilions' incendiary films all intimately resembled one another, even in their duration. An identicality and repetition inhabited those artists' films of misfired disintegration, just as the corporate pavilions' architectural format – all with their multiple projectors and large-scale immersive screens, installed in plastic-faced domes – followed an intractable regime of identicality and repetition, as though driven into existence by Muybridge's own obsessions with repetition and his incorporation into his work of malfunction. The contents of the pavilions' films appear rarely to have troubled their commissioning corporations, whose unique concern was for their pavilions to be technologically innovative.

As the audiences emerged from the exits through which they were expulsed, clearly exhausted and sweat-drenched from their experiences at the hands of the pavilions' unseen projectionists, their responses were filmed for the corporate documentaries. Many spectators hurry by fearfully, but others pause, smile in relief and tell the interviewer that, over the past minutes, they have witnessed something unprecedented, and have experienced sensations never-experienced before; that projection has astonished them. Their still-deranged eyes appear enduringly illuminated by cataclysmic fire, as though holding the retinal images that Muybridge's contemporary, the ocular anatomist Wilhelm Kühne, had aspired to record. In the background, behind the interviewees, new contingents of spectators are already streaming inside for their projection experience. Especially, young spectators are interviewed, and declare simply that their experience was great, amazing, as the audiences of Muybridge's 1891 Central European projections also had.

Once the last of its 64,000,000 visitors had left that experimental city in September 1970, the Osaka Exposition's terrain of immersive moving-image pavilions was razed, as thoroughly as Muybridge's Zoopraxographical Hall – already usurped by Pandolfelli's Pompeii Panorama – and all of the other Chicago Exposition Midway Plaisance attractions and main-site pavilions had been destroyed, on their closure. Whereas the Chicago Exposition's authorities themselves supervised the fires and demolitions which transformed that site into a scorched-earth expanse, each of the corporate directors of the Osaka Exposition's pavilions was contractually obliged to ensure their vanishing occurred within a span of six months. Several of the more adroit corporations managed to sell their pavilions' carapaces for dismantlement and re-installation in distant public parks, but most were simply flattened into their constituent panels, then incinerated or buried under their sites (Bini's domed pavilions were designed to be destroyed in the same brief interval that it took to inflate and install them). By the beginning of 1971, aerial photographs of the site showed an apocalyptic wasteland, still striated by the network of roads that had led visitors between the pavilions, and closely replicating the 1894 photograph "World's Fair Chicago 1893 Wrecked after Closing", as though the entire Exposition site had been created in order for its razing to form a re-enactment of that photograph. Only one pavilion, the Electricity pavilion, was retained, along with the Tower of the Sun and a lake, constellated by water-sculptures designed by Isamu Noguchi, in which children had swum in respite from their visits' exacting moving-image projection experiences. A semi-wild park was eventually installed on the Exposition's site.

In future decades, further Expositions were constructed worldwide and then promptly demolished. The Japanese photographer Ryuji Miyamoto photographed part-demolished projection-auditoria in Tokyo's Asakusa and other districts, as the city's cinemas fell into decline, and also photographed the derelict site and torn-down pavilions of the 1985 Exposition in Tsukuba, a far smaller-scale event than the Osaka Exposition, held at a purpose-built "city of technology" located to the north of Tokyo (a further, intermediate Exposition had taken place in 1975, on Japan's southern island of Okinawa, focused on the exploitation of oceanic environments). Miyamoto collected his photographs, oscillating between the demolitions of cinematic projection sites and the Tsukuba Exposition site, under the title *Architectural Apocalypse*. The Tsukuba Exposition's site, as with the Chicago Exposition's Midway Plaisance and the Osaka Exposition's Expo Land, pivoted around the installation of an immense Ferris wheel, but it no longer manifested the driving preoccupation with experimental moving-image projection of the Osaka Exposition, focusing instead on exterior-space urban digital moving-image screen projections and on new domestic technologies, as though the event were designed as an after-flicker or spectral shadow of the Osaka Exposition's moving-image obsessions. Miyamoto's photographs show ripped-apart pavilions in triangular rather than domed shapes. Miyamoto always noted the dates of the construction and ruination of the cinemas and pavilions that he photographed, with the cinemas usually holding a duration of sixty years or so; the immediacy of the Tsukuba Exposition site's razing – its pavilions built in 1985, then demolished in 1985 – indicated an accelerating urgency in the erasure of Expositions' traces (evident too in later expositions such as that held in Shanghai in 2010, which finally exceeded the Osaka Exposition's total of visitors, and impending expositions, as in Osaka's 2025 future-repetition of its 1970 Exposition).

The photographic historian Mark Holborn assembles the far-flung urban elements of Miyamoto's work around a pivotal preoccupation with ruination: "Miyamoto's ruins, especially those of cinemas and theatres, or most conspicuously those of the Tsukuba Exposition site from 1985, where the future of Japanese technology was presented, are relics from lost cities now surpassed not by the ruins of fabric and matter but by the ruins of technology."[5] But equally, or more so, Miyamoto's work is focused on the detrita of projection, and on the resilience of projection, even when projection appears to be in the process of obliteration or to have disappeared altogether. Miyamoto's work could best be seen in anonymity, with his photographic sequences of the Tsukuba Exposition appearing as a further accentuation in the relentless repetition of the erasures of Expositions as sites of projection experiments, manifested most formatively by the Chicago Exposition's demolition (encompassing Muybridge's projection auditorium) and by the wasteland-generating mass-razing of the Osaka Exposition's many projection pavilions, with both of those seminal events documented by photographers whose names are unknown, as with the names of Muybridge's touring moving-image projectionists.

The corporate directors of the Osaka Exposition's pavilions had no further interest whatsoever in the experimental projectors which had inhabited their auditoria, nor in the artists' films which those projectors had transmitted onto the pavilions' immersive screens. The projectors' exorbitant dimensions and gauges were too unwieldy for commercial adaptation in cinemas, and the artists' films – watched by audiences in their many millions during the span of the Exposition – were rendered instantly obsolete on its closure; many films were discarded by their corporate proprietors and lost altogether, or survive only in brief fragments through their sequences' mid-projection filming in the corporate documentaries shot inside the pavilions, as with those commissioned by the Toshiba and Midori-kai corporations. The projectors, as moving-image technology's redundant detrita, were either destroyed or stored-away in oblivion, or else given away to the Exposition's Documentation Center, established soon after the event's closure in the three storeys of an adjacent administration building, where they were assembled in a large storage area on the uppermost storey.

All of the Osaka Exposition's many hundreds of projectionists were also rendered redundant on the day of its closure. Since many of them had also been participants of the film-crews which had shot footage on location for the pavilions' projections, they had often been engaged in the Exposition project for two or three years. The technical complexity of the immersive projections had demanded extreme, unrelenting concentration from their projectionists; the films were projected many times each day, each fifteen or twenty minutes, with no respite. Since the Exposition's opening hours had been long, extending from 9am to 10.30pm for seven days a week during the summer months, the projectionists had been expected to exhibit exceptional alertness, often sleeping in their projection-boxes after their seventeen-hour shifts (the films needed to be meticulously checked for rips and defects, before and after each day's projections) and never leaving the site for months on end. At the Exposition's closure, the projectionists' bodies themselves formed corporeal wastelands that doubled the site's own razing. Since the Exposition projections constituted an aberrant one-off phenomenon and their corporate directors viewed any sequel to them as being worthless, the projectionists were simply jettisoned from the site on the event's last night, abandoning their projectors, still burning hot.

I visited the Osaka Exposition's Documentation Center and met its current curator and its original curator from the beginning of the 1970s, now elderly and retired but still avidly compelled to remember the event and to show me whatever I wanted to see. After some weeks there, I asked to see the multitude of abandoned projectors on the building's uppermost storey. We ascended in the decrepit elevator of that never-restored building, which still emanated intact the exhilaration and abrupt erasure of the 1970 event. The retired curator found the key and opened the door. Inside that space, through a pall of dust, I saw the projectors – many of them in identical form, since their pavilions had needed five or more synchronised projectors to animate an immense immersive screen – extending far across the linoleum-floored storage space. It felt as though a roomful of the figures of long-term detainees, confined to indefinite darkness, had abruptly been exposed to light by their imprisoners. The projectors had all been separated from their film-reels, but were otherwise intact, ready to be resuscitated. The 1970s curator told me that he was now exhausted and had to go and sit down in the lounge-area situated on the storey below, and handed me the key. I remained there for one hour, in company with the amassed projectors of the Osaka Exposition, in their profound abandonment, then locked the door and rejoined him.

The 1970s curator was gazing out of the lounge's plate-glass window that overlooked the vast parkland now occupying the Exposition site, quietly weeping. I asked him if he knew of any surviving projectionists from the era of the Exposition, and after some hesitation, he pointed towards the park below and told me there was one, named Seiji Mochizuki, who spent all of his days circuiting that park below, still obsessed with his memories of his work as an experimental moving-image projectionist. If I wanted, he could call him now; a moment later, I saw the crumpled face of the ex-projectionist on the iPhone's screen, assenting to a meeting, and a few minutes later, I had entered the park and met him outside the last-surviving pavilion, the Electricity pavilion. I took a long walk around the park with that haunted, still-itinerant ex-projectionist, who was immediately ready to talk. First of all, he told me that he had also been an Exposition cinematographer, as well as a projectionist, and had travelled throughout 1968 and 1969 to the innumerable locations at which the film was shot for the under-construction pavilion whose corporation had employed him. The stormy sea-crossing between the ports of Oma and Hakodate, to film on the volcanic terrain of the island of Hokkaido, had been so rough that he and his colleagues had needed to put their arms around the film-cameras to prevent them being blown into the ocean. They had travelled outside Japan too, always itinerant, to San Francisco, and other cities, he couldn't remember them all. The film they had shot was a strange one, he could never understand it... We began to circuit the deserted lake on which Noguchi's now-derelict sculptures had once blown great spouts of water onto the ecstatic children swimming below them. Now, the ex-projectionist spoke of his childhood in Osaka, during which he had somehow been ostracised and prevented from joining the usual junior-level gangster-clans, though he had committed many youthful crimes, of course. He had found a job as a projectionist by accident at the beginning of the 1960s, in a city-centre alleyway cinema, then had projected horror and pornographic films through most of that decade, soon habituated to acute solitude in his white-hot, cramped projection box, while that decade's urban-Osaka riots flared in the streets outside, sometimes even audible in the cinema itself.

When he had heard that the Exposition's corporations needed to hire hundreds of projectionists, and would train them for experimental projections, he had applied. He explained: "My parents were from the *burakumin* districts of Osaka: do you know what that is? – the skinners, slaughterhouse-workers, transporters of the body's wastes. Osaka's outcasts. But with the Exposition projectionists, with my team of projectionists, I was together with them immediately, for the duration of the Exposition... All projectionists are outcasts. The work was hard, so many hours each day. But we always kept going somehow, we never had a breakdown. I slept in the auditorium or in one of the projection-boxes, as we all did, always exhausted, or walked through the Exposition grounds at night." At a certain point during the span of the Exposition, he had transferred from one pavilion's projection staff to another, for reasons he declined to explain. We walked over a large, featureless picnic lawn on the edge of the park's wild zone, and the ex-projectionist suddenly stopped and pointed at the ground: "Here was the Toshiba IHI pavilion auditorium." A few minutes later, again over unmarked grassland, he stopped and pointed downwards: "And here was the Midori pavilion auditorium."

I asked him what had happened when the Exposition closed. We were now standing by the base of the Tower of the Sun, close to the park's exit. He smiled and replied: "Then I returned to my solitude." I asked if he had gone back to the white-hot projection box in urban Osaka, but he was now withdrawn and said nothing further, motioning for me to exit the park so that he could restart his circuiting itinerance around its edges, to generate oblivion towards the contemporary moment and to re-immerse himself in his time as a projectionist of the Exposition. I asked him if any of his fellow projectionists were still alive, but he stared at the ground and simply murmured: "All dead now, all dead." I decided to attempt one final question, and asked Mochizuki-san if he knew anything of Eadweard Muybridge's moving-image projections in 1894 Japan, and now his eyes were instantly alert and activated again. No, he knew nothing of that, but he was eager to declare his great regard for Muybridge: "In 1970, our two most respectable men in Japan: Eadweard Muybridge and Jean Genet... you know our beloved criminal of film, 'Saint Genet'?... Perhaps 'Saint Muybridge' too, yes?... Muybridge, the killer...". He walked away.

Notes

1. *Japan World Exposition, Osaka, 1970: Official Report* (in three volumes, colour-coded Blue, Red and Green), Commemorative Association for the Japan World Exposition (1970), Osaka, 1971, page 11 of Red volume. This chapter was primarily researched at the Documentation Center of the Osaka Exposition, located next to the former grounds of the Exposition and its "Expo Land" annex; I'm very grateful to the Center's current archival director Masashi Tatsukawa and to its 1970s archivist Toshihiku Okaue who, though long-retired, met me there to discuss his experiences of the Exposition's pavilions and to show me conserved projection artefacts. I'm also grateful to Yu Homma for arranging the complicated permissions for me to visit the Documentation Center, and also to the many people in Japan and Europe who spoke to me about their attendance, often as children, at the Exposition.

2. Ibid, page 375 of Blue volume.

3. Interview with Dante Bini, Arezzo, 24 June 2016.

4. *Japan World Exposition, Osaka, 1970: Official Report*, pages 422, 433 and 432, and the Toshiba IHI pavilion's "Global Vision Theater" programme, Toshiba Corporation, Tokyo, 1970, which states its anodyne corporate goal on page 8: "Let's all join, soul and smile/wherever we go, and whatever we do."

5. Mark Holborn, *Beyond Japan: A Photo Theatre*, Jonathan Cape, London, 1991, page 39.

Lovers: Teiji Furuhashi's Projections, 1994/2016/2020

Alongside the inspiration of Muybridge's projection experiments – undertaken as a touring and fixed-space projectionist, and as the moving-image projection auditorium's innovator – for immense experimental urban spectacles for public audiences, as at the 1970 Osaka Exposition, that inspiration also vitally infiltrated the work of individual artists, whether allied to corporations or working autonomously, and does so still more urgently now, in contemporary art and urban environments. Muybridge's inspiration, as with his work's own conception and its projections, integrally forms an irregular entity: non-linear, mysterious, spectral, contrary, malfunctional. It can also incite a "haunting", to use the term by which Francis Bacon invoked Muybridge's inspiration for him: a haunting which is propelled backwards by Bacon towards Muybridge's sequences through Bacon's underfoot trampling, violent smearing and laceration of photographs torn from the multiple copies he owned of *The Human Figure in Motion*, as a preparation for Bacon's subsequent autonomous rendering of Muybridge's images, as a form of exacting corporeal projection.[1] While Muybridge often incorporated his malfunctioned images into his sequences, thereby erasing any narrational resonance in them, Bacon habitually destroyed those of his works which he considered to be misfired, generating a streamlined density of sequential repetition for his body of work, as with the characteristic approach of experimental film-artists attempting to avoid any trace of narrative in their work, at all costs.

Muybridge's inspiration for individual artists was transformed by the ascendancy of the digital image across the first half of the 1990s, in a parallel way to that in which digital projection also rapidly began to supplant celluloid projection – and transformed, too, the work of the moving-image's projectionists – in that era, which is now itself one of an obsolescence that appears far more acute and irreparable than the era of Muybridge's own technologies of projection. The residues of Muybridge's work – his photographic sequences, his Zoopraxiscope projector and projection discs, and his self-monumentalising scrapbook and other documents, along with his creative and corporeal aura of unrelenting excess in undertaking that work – generated a multiple inspiration in digital film-art experiments from the 1990s on, as well as in the time-mutated dimensions of motion-capture and other sequential technologies that also soon permeated studio-film productions, most notoriously via *The Matrix* films. Contemporary motion-capture studios, such as Andy Serkis's Imaginarium Studios, form direct successors, as corporeally-focused experimental moving-image laboratories, to Muybridge's Zoopraxographical Hall. But above all, Muybridge's work is deeply embedded in the experiments of artists engaged with the status and future of moving-image projection.

One of the earliest and most pivotal of those digital-projection experiments, the artist Teiji Furuhashi's work *Lovers*, was undertaken in Japan in 1994, at the sensitised, volatile interval between the expiry of video and celluloid projection media, and the ascendancy of digital media. Furuhashi's work is also illuminating for its re-conception of the status of the moving-image projectionist, at that transitional moment of projection's upending. *Lovers* was screen-projected in relatively small-scale gallery auditoria as an immersive environment for its spectators, occupying spatial dimensions (usually, thirty-two feet by thirty-two feet) far closer to those of Muybridge's Zoopraxographical Hall than to those of the pavilions of the Osaka Exposition; even so, *Lovers* combines the extreme innovation in exploring projection's capacities which both of those previous experiments had embodied. While Muybridge's moving-image projector, his "operator" and his own commentating body were all simultaneously exposed in his projection space, the Osaka Exposition's projectionists and projectors were concealed in their projection-boxes, with only the pavilions' immersive screens visible. For *Lovers*, Furuhashi revealed his projection media, as Muybridge had, in the form of a metal-framed tower installed prominently in the space (the tower held a part-digital, part-analogue content: five computer-controlled moving-image projectors, each playing laser discs, and two 35mm-format slide-projectors, with all seven devices constantly rotating), but as his spectacle's own projectionist, Furuhashi had vanished, his figure now transplanted into his moving-image screen.[2]

Lovers' projections are drawn directly from Muybridge's Philadelphia sequences of walking figures and of figures abruptly self-arrested, sharply turning as though convulsed, in order to retrace their walk's trajectory or else to break into a run; Muybridge incorporated many of those sequences into his projections on his Central European tour and at his Chicago projection auditorium. On *Lovers'* immersive screen, enclosing the darkened projection space and its standing spectators, five young, naked figures appear to be predominantly in transit. In their walking, those figures occasionally cross over each others' paths – sometimes moving steadily, sometimes in abrupt acceleration – and those figures then become transparently layered over one another, while each figure remains autonomous, never interconnecting with or acknowledging the other figures. The figures' bodies occasionally come to a twisted halt as though frozen, in the form of a danced-out corporeal malfunction, arms extended horizontally across their own waists and throats, apparently vulnerable to erasure in that stasis; then they begin walking again, or burst into runs. The projection space's spectators also stand in stasis, or else walk to track or replicate the projected bodies' movements, their ocular attention divided by texts imprinted on the space's floor, adapted from writings by Jean Genet and focused on corporeal boundaries and border-crossings, as though forming a succinct commentary on the projections, or more widely upon controlled societies, their perimeters tightened still further in Japan from the 1960s and Osaka Exposition eras by digital surveillance media. If left to run without interruption, *Lovers* lasted for around fifteen minutes, as with many of the Osaka Exposition's moving-image experiments.

Furuhashi's own naked figure is one of the five figures inhabiting *Lovers*' immersive screen. As well as being the filmmaker who has already captured his work's corporeal sequences of movement, he is also the body in movement that is enclosed within those sequences, and is additionally his own projectionist, who no longer needs to stand directly beside his tower of projectors or to even appoint a surrogate operator to do so. The projectors' digital technology enables them to operate without human intervention, and so Furuhashi is able to immerse himself entirely in his projections. As a result, the projectionist enters into the screen, and is now beyond any need to vocally commentate on his act. But once installed in his screen, that projectionist is cast into a still more acute vulnerability than the other four projected figures. In *Lovers*, Furuhashi's figure is the only body that will tilt vertiginously backwards, and plummet into its own vanishing, whenever a spectator stands directly in front of it, thereby ending the projection. Muybridge, too, had incorporated his own naked body into his sequences (a far older body than Furuhashi's, its face bearded, its gait heavier), alongside those sequences' younger figures of Philadelphia athletes and students; that elderly projectionist's body, held within the screen, always accomplishes its actions before elapsing. By contrast, Furuhashi's isolated projectionist's body, transposed into his screen, will aberrantly dissolve at any moment into darkness. *Lovers* holds infinite repetition within its sequences, through their permutations of movements as well as those movements' readiness for malfunction, in ways that are allied to Muybridge's own driven preoccupations with misfired repetition.

Furuhashi's *Lovers* was his sole experiment as an individual artist in creating a moving-image projection installation and embedding his own figure within it, outside of his collectively undertaken work as a participant of the prominent Kyoto-based performance-art group Dumb Type; he died from AIDS-related illnesses in the following year, 1995. He said almost nothing in the form of commentary on *Lovers*, declining to explicate it in a vocal dialogue with the curator Barbara London by repeatedly declaring: "I don't get the question", and thereby reinforcing the work's isolation. Both in its isolation as an experimental artwork and in its conjuration of a corporeal isolation that can imminently lead to an enforced vanishing, *Lovers* is intimately close to, and conjoined with, Muybridge's own projection experiments, projected in Central Europe's cities and in Chicago, exactly a century earlier. Similarly, the Osaka Exposition formed the isolated one-off instance of an immense-scale public event devoted primarily to multiple spatially-located experimental moving-image projections of the kind that Muybridge had himself undertaken only once, in one venue, in Chicago. Isolated and unprecedented experiments in projection may ultimately come to form strange and unanticipated amalgams over time, in a parallel way to that in which projectionists – habitually isolated in their projection-boxes, and often hostile to their obsessional work's interruption – can contrarily ally with one another to form temporary teams, as with those combined to operate the Osaka Exposition pavilions' multiple projectors and also to shoot the films projected there, as cinematographic projectionists. As with those of *Lovers*, projectionists' bodies intersect and overlayer one another in time, but form spectral entities that never touch.

As a Muybridge-inspired digital-projection work incorporating the living body of its projectionist, *Lovers* only existed for one year, from 1994 to 1995; after that, with the death of its projectionist creator, only the artwork's raw mixture of technological traces remained, and the spectrality of the body which it enduringly held – in anticipation of future projections – became accentuated still further. *Lovers* was the work of an individual artist "haunted" by Muybridge's work in Furuhashi's corporeal experimentation with projection, but it was never an autonomous work, and remained permeated by its origination in a corporation's commissioning process, as with all of the artists' moving-image works that had been projected within the Osaka Exposition's corporate pavilions, a quarter-century earlier. At first sight, *Lovers* appears to form the isolated antithesis to that Exposition's profligate scale and to the profound corporate infiltration of the art-film projections in its many pavilions, but Furuhashi's work operated, even in its relatively miniaturised form, in intimate alliance also with those dynamics of corporate image-regimes.

The media corporation Canon Inc., which had originated in the 1930s as a small-scale company designing and manufacturing photographic cameras, operated an experimental annex in 1994, known as "Canon ArtLab", which commissioned artists' moving-image works with an emphasis on immersive digital projections. Sometimes, Canon ArtLab short-circuited its work with individual artists, and instead conceived of immersive or virtual-reality installations generated around cryptic "themes", paralleling those envisaged by Osaka Exposition corporations such as Toshiba. With commissions such as Furuhashi's *Lovers*, the moving-image artists developed their own independent preoccupations (as with *Lovers*' emphasis on projections of isolated, death-striated corporeality in movement) without corporate intervention; Canon ArtLab then funded *Lovers*' development and devised the projection technologies that would activate it. That oscillation between the artwork's independent conception and its corporate realization renders *Lovers* a volatile and exposed entity, distinctly unique but non-autonomous; in that sense, it resonates with the precarity of Muybridge's unprecedented projections at his self-instigated Zoopraxographical Hall, realizable only through the Chicago Exposition authorities' arbitrary granting to him of a "concession", and also with the vulnerability of the artists' films projected immersively for audiences of millions at the Osaka Exposition, as the commissioned (and readily discardable) property of the corporations which had also commissioned the pavilions in which those works were projected.

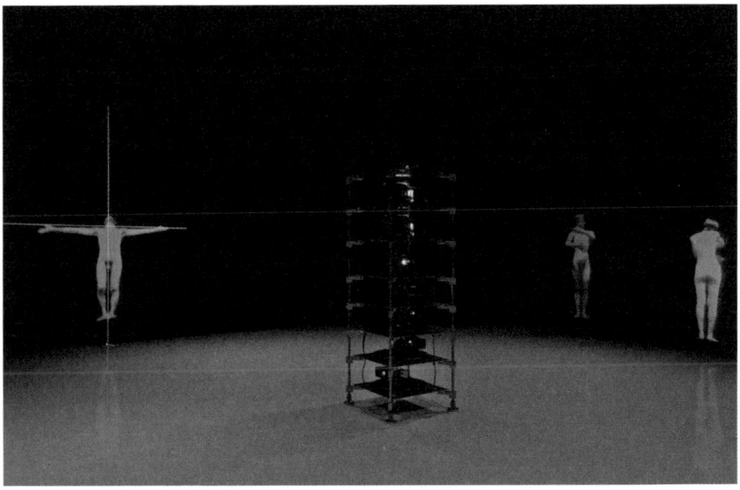

Ill. 22: *Lovers* with its tower of projectors and Teiji Furuhashi's falling body (on the left-hand side), Tokyo, 1998.

Lovers was presented for the first time, in September to October 1994, using an immersive projection screen, at the Hillside Plaza gallery in Tokyo's Daikanyama district; as well as displaying Furuhashi's digital moving-image artwork, that installation simultaneously showcased the adept technological innovations of Canon ArtLab in intersecting new digital projection media with now-vanishing, obsolete analogue media. The Hillside Plaza formed a complex of showrooms and studios with its elongated, low-ceilinged art-space's entrance on the ground floor; spectators entering the darkened space were guided by the presence of uniformed hostesses, as happened too with the disorientating pavilions of the Osaka Exposition. Following Furuhashi's death, Canon ArtLab again displayed *Lovers*' projection installation in May 1998 at a more prominent Tokyo venue, Spiral Hall, though still in its concentrated dimensions which required its space's spectator-occupants to be limited in number in order not to collide in darkness or to precipitate excessive vanishings of Furuhashi's figure through the repeated act of standing in front of it. *Lovers* was projected at other venues in that era, as at the New York Museum of Modern Art in 1995.

To promote its support of digital artists and its own technological expertise, Canon ArtLab gave away entire sets of the projection technology of *Lovers* – encompassing multiple editions of its metal projection-tower, projectors and digital projection-data – to several of the venues in which it was shown, including to the New York Museum of Modern Art, in 1998, as though Canon ArtLab operated as the sole proprietor of *Lovers*' potentially expansive future acts of projection as well as being the artwork's commissioning instigator. That pivotal loss of Furuhashi's status as the work's filmer and projectionist (a projectionist inhabiting his own work's screen, and vanishing from it) was only enhanced through his death in the period preceding those donations. After *Lovers*' Tokyo projection in 1998, the work faded from sight. Canon ArtLab abruptly ceased to exist in 2001 after Canon Inc. annulled it following curatorial disputes, as though that corporation had become exhausted already by the exacting spectatorial demands of the preceding years' digital-art projections; it now invalidated and deleted all future imperatives to display its ArtLab annex's innovations.

As a result, *Lovers* acquired the status of a seminal but lost work of projection across the 2000s and most of the 2010s, its digital data submerging from sight, as had happened with the textual data of press-releases and press accounts that had documented Muybridge's Central European projection tour and Zoopraxographical Hall projections, and with the Osaka Exposition's printed promotional programmes, once those events' live-instant immediacy of projection in their auditoria had disintegrated. As a last-ditch, last-gasp moving-image project undertaken at the extreme end of Furuhashi's work, *Lovers* became stranded in time, detached also from his earlier works – undertaken with his activist collaborators in the Dumb Type performance-art group – which had more explicitly combatted Japan's systematic neglect of AIDS patients in the years before his death, or had attempted to confront other governmental and corporate failings. That body of work of Furuhashi in collaboration with the Dumb Type group, also vitally inspired by Muybridge's work, generated its own distinctive inspirations for future digital artists' projects with ecological or apocalyptic preoccupations, or with the desire to interrogate the controlling powers of technologies, or aiming to experiment with urban corporeality in intimacy with digital image-making. But Furuhashi's *Lovers*, with its particular obsessions located in projection and the screen-installed, isolated figure of the projectionist, fell into oblivion.

By 2016, when the New York Museum of Modern Art's curators became interested in resuscitating *Lovers*' projections, more as a technological challenge than to re-activate Furuhashi's Muybridge-inspired figures in movement and convulsion, its projectors' deep descent into obsolescence and abandonment required an exacting excavation into the early era of art's digital moving-image installations. The Museum's curators extracted all of the projection material donated to them by Canon ArtLab after *Lovers* had been installed in one of the Museum's galleries, from the storage facilities in which it had been left in suspension, and examined that material with the aim to determine whether it could be assembled for a new projection of *Lovers*. Soon, they determined that it could not, but the long-expired projection technology could instead be supplanted and upgraded, to generate a new variant or entity of *Lovers* for projection, standardised with contemporary digital data systems. With the work's corporate commissioning body rendered defunct, and Furuhashi's own death, there emerged an unconstrained institutional freedom to re-envision and overhaul the work, and to synchronise it with the then-current technological context. *Lovers* would need to be transformed in order to appear identical to itself, in its 1994-98 projections, but at the same time, its constituent digital anatomy would become utterly other.

The Museum's specialist art-technologies curator and archivist, Ben Fino-Radin, compiled a report on *Lovers*' upgrading, from the moment when the curators, with bewildered horror, extracted the malfunctioned and incomprehensible data-files and projectors from their storage areas, to the contrary moment of achievement when *Lovers* was revived and opened for its inhabitation by new spectators from July 2016 to April 2017 (a far longer duration than *Lovers*' first Tokyo projections at the Hillside Plaza gallery and Spiral Hall, which lasted each time only for several weeks). That accomplished reactivation of a lost projection event – and of the projectionist figure of Furuhashi, still transposed from the projectors' tower into *Lovers*' screen, then propelled into its vanishing, many times each day – invokes the potential also to reanimate Muybridge's own 1891-93 projection events, in their locations, as at Berlin's Urania auditorium, or even to dig the ground of Chicago's Midway Plaisance where the wrecked structure of the Zoopraxographical Hall may be interred, to reassemble it, since Muybridge's Zoopraxiscope projector and discs survived intact his moving-image projectionist's itinerance and only require the restitution of their authentic spaces of projection.[3]

In defiance of the *Lovers* report's sense of technological triumph after four months of painstaking restoration ("High fives were exchanged" on the work's completion, as Fino-Radin notes), that report is also contrarily striated with the sense that Furuhashi's 1994 work, in its reconfiguration, has been voided into a mutation of itself that cannot be sustained into the future. Notably, all of *Lovers*' original projectors became jettisoned. Fino-Radin concludes his report: "Just as the original equipment that controlled *Lovers* had aged, obsolesced and become unusable, so will our newly restored solutions... The day may come when our work here seems somehow wrong or misguided...". No commentaries of Furuhashi himself could be deployed to justify the work's upgrade, since he had been a resolutely near-silent projectionist of *Lovers*, in an inverse form to Muybridge's status as an excessively voluble, physically present commentator of his own projections. Even so, that exacting institutional work of upgrading had been carried through to its end, through a driving compulsion to reanimate *Lovers*' projections via the technological and sensory imperatives of the contemporary moment. Fino-Radin emphasises that element of urgent necessity, suggesting that if a work such as *Lovers* remained perpetually confined to an art museum's archives, never to be brought out again for new projections, it would eventually disappear or die completely. The obsolescence of the early era of digital moving-image projection, in the first half of the 1990s, appeared infinitely deeper and more intractable than that, for example, of Muybridge's era of projection, in part through 1990s artists' frequent use of digital technologies in contaminated mixings with previous, already-redundant technologies, as with *Lovers*; Muybridge, by contrast, worked with almost nothing that was pre-existent, generating moving-image projection in unprecedented autonomy.

The then-contemporary technologies of the moment of *Lovers*' upgrade are themselves already obsolete, only a few years later, through digital technologies' wildly accelerating mutations, potentially appearing still more profoundly archaic and dysfunctional in the contemporary moment than *Lovers*' mid-1990s corrupted amalgam of technologies had appeared in 2016. *Lovers*' mixings of digital and analogue formats had been viewed as adroit creative approaches in their own era, but for Fino-Radin, they had come to generate impasses that needed to be entirely excised, rather than adapted, in the process of a "migration" of projection data that necessarily also entailed the post-death migrancy of Furuhashi's already-itinerant projectionist's body. Fino-Radin notes: "The original LCD video projectors and the behind-the-scenes control hardware needed to be replaced due to their instability and rarity. This meant a full-on re-implementation of the original control and timing hardware and software would be necessary".[4]

The restored institutional version of *Lovers* projected in standardised digital formats at New York's Museum of Modern

Art galleries in 2016-17, following its corporate projections in Tokyo of 1994-98 overseen by Canon Inc.'s ArtLab, was fine-tuned and endorsed in its last stages by two of Furuhashi's Dumb Type-group collaborators, Yoko Takatani and Shiro Takatani. That reactivation of *Lovers* conjured new, ineradicable corporealities that form an antithesis to the more erratic, flaw-embedded bodies, exposed and attuned to disappearance, of its earlier projections, which resonate more closely with Muybridge's own Philadelphia corporeal sequences and their projections on his Central European tour and at the Chicago Exposition, incorporating malfunction and mishap, alongside their propulsive beguilings of their spectators' eyes. An alternative option to that institutional approach of rendering *Lovers* identical with contemporary digital image-regimes – by contrast to the mid-1990s era of experimentation, those regimes were overwhelmingly boundaried within the domain of corporate power by 2016 – would have been to exhibit and project *Lovers* in its time-exposed state of obsolescence, malfunction and disintegration, even in a condition beyond projection, as projection's final breakdown and failure. *Lovers* is, after all, a work that projects the fall into death. The awry, zigzagging historical lineage of experimental moving-image projection's many expirings and vanishings – via the originating closure and destruction of Muybridge's Zoopraxographical Hall, the detonation of the Skladanowsky brothers' Central-Hallen projection-auditorium and the razing of the Osaka Exposition pavilions – forms an integral presence within *Lovers*, extending from its spectators' clear view of the projectionist-vacated tower of projectors as the pivotal element of the auditorium, to their own capacity, as spectators standing directly in front of Furuhashi's figure, to enact an ocular and corporeal "projecting" of that exposed figure into its backwards-fall and disappearance from its screen.

Tokyo itself is a megalopolis of ecstatic falls and disappearances, including the vanished traces of Muybridge's self-declared visit there in 1894. From the archives of the Tokyo Photographic Art

Museum, where I had been scouring the records of that year's photographic society meetings for any mention of Muybridge appearing with his projector in Japan to give a demonstration of his work, and finding nothing, I ascended the hill to the salubrious Daikanyama district of art bookshops and embassies. Since the first projection of *Lovers* at the Hillside Plaza gallery in 1994, exactly a century after Muybridge's own phantom projections in Japan, that venue had been adapted into a hall for corporate receptions, following the 2010s entrenched divergence of corporate digital media from experimental art and projection (a divergence prefigured by Canon Inc.'s closure of its ArtLab in 2001). I knew that, on a Friday evening, a lavish corporate reception would certainly be held in that space, celebrating a new project of Toshiba, Fuji Group or Midori-kai, a vast central table of sashimi and sake occupying the site of Furuhashi's tower of projectors. If I could infiltrate that reception, I would be able to assess any residual presence of *Lovers*' projections of a quarter-century earlier, in its flooring perhaps still scraped by that tower's metal base or with its Genet-texts inadequately erased, a fragment of their boundary-denouncing lettering still imprinted under the executives' shoes. Even the velocity of Furuhashi's backwards-plummeting into void projection-screen space could conceivably leave a marker of some kind, since the work of a projectionist will vanish, but is always ineradicably embedded within the sites it inhabited.

When I reached the Hillside Plaza building, I found that the entire complex was being demolished, and had already been shattered and gutted, though the demolition crew had evidently finished their work for the evening and would only complete it on the next day. Several uniformed security-guards were lined-up in front of the taped-off entrance to the Hillside Plaza gallery, where uniformed hostesses had once stood at the moment of *Lovers*' projections, to assist spectators blinded by the space's darkness, only intermittently illuminated by Furuhashi's Muybridge-inspired figures in movement. I had to wait for two hours until those guards took a break, which I knew would only be for a few seconds, since they were alert to any intrusion, holding illuminated sticks in their white-gloved hands, but I thought I could now enter the building, even momentarily. At least the building's surveillance cameras had already themselves been demolished. After I had slipped-through the taped-off entrance, I saw how closely the near-demolished interior resembled one of Miyamoto's zones of "architectural apocalypse", the walls buckled and shattered, the floor voided, the ceiling part-collapsed. The bulldozers had entered the space through a rear side-wall, hidden from the main Daikanyama avenue. Those bulldozers had both excoriated and torn-open the core of that projection auditorium, abruptly exposing its interior – and any obstinate detrita of *Lovers*' bodies – as though through an act of anatomical projection. I stayed only for an instant, and returned the following evening to find the entire Hillside Plaza complex flattened, most of the flimsy 1980s building-materials already promptly removed from the site, so that it now formed a wasteland in the heart of Tokyo, as with the many wastelands of the Chicago Exposition site including that formerly occupied by Muybridge's Zoopraxographical Hall, "wrecked after closing", of the Skladanowskys' Central-Hallen auditorium site, its debris heaped into a burnt-out line, and of the Osaka Exposition pavilions' sites, their domed carapaces urgently crushed and ready to be buried underfoot. The work of the projectionists inhabits a still-animated, multiple wasteland of images, bodies and cities.

Notes

1. David Sylvester, *The Brutality of Fact: Interviews with Francis Bacon*, Thames and Hudson, London, 1987, page 30.

2. This chapter was researched at the archive at Kyoto's Seika University which contains materials surrounding *Lovers*. It draws too on my experiences of inhabiting *Lovers*' projection space at Spiral Hall, Tokyo, in 1998, and at the Museum of Modern Art, New York, in 2017.

3. If Muybridge's projections, from his Central European tour or at the Chicago Exposition, were to be future-reanimated, in their original sites such as Berlin's now-abandoned Urania auditorium or on the Chicago Exposition's Midway Plaisance avenue, it would be necessary to position his multiply-layered, self-monumentalising scrapbook of 1894-1904 directly alongside the projection screen, as the closest equivalent to his vocal live-moment commentary, often pursued at excessive length.

4. Ben Fino-Radin, "Art In the Age of Obsolescence: Rescuing an Artwork from Crumbling Technologies", website of the New York Museum of Modern Art, 21 December 2016. Fino-Radin also analysed the dilemmas of reconfiguring *Lovers* in a 2016 conversation with the art-writer Christopher Schreck, also accessible online.

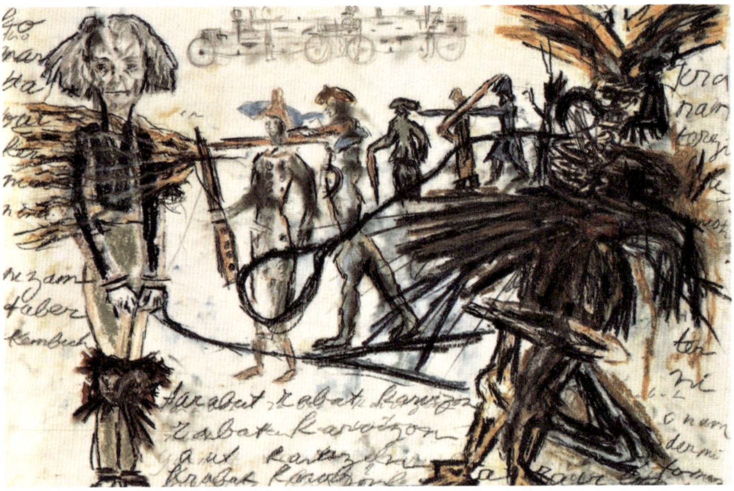

Ill. 23: Antonin Artaud, *The Projection of the authentic body*, 1946-48.

Projections of Future Cities, Projections of Future Bodies

At the beginning of 1948, shortly before his death, the artist and theorist Antonin Artaud completed a drawing entitled *The Projection of the authentic body*. Work on the drawing had begun in an insane asylum, two years earlier, but was finished only after Artaud's release and his return to an autonomous existence in Paris. On the left-hand side of the drawing, Artaud's figure is under attack from several uniformed assassins, one of whom fires a rifle that unleashes multiple streams of orange fire across Artaud's upper body. His kneecaps have already been assaulted and appear to have been torn-open and exposed to the bone. Artaud's face is staring directly at the drawing's spectator. His wrists have been bound, but rather than incapacitating him, the bindings lead on to a rope which loops and intimately attaches Artaud's figure to the waist of another figure, which occupies the right-hand side of the drawing. That second figure, Artaud's envisioned double, is not under attack, and itself propels thick projections of fire from the torso of its feral, skeletal form; that retributory projection extends across the uniformed figures as though about to massacre them. That double's entire upper head, too, is emitting flames, forming a kind of incendiary headdress. The face of Artaud's second, projected body also stares directly at the spectator, and though that body is tilted obliquely, it appears far readier to endanger its spectator, through the wild capacity for projection which Artaud instils in it, than to fall. Its mutation into a more "authentic" corporeal duplication propels Artaud's body out of its space of assailed subjugation, and the corporeality that is unleashed by the act of projection is too insurgent to be installed in any future's fixed or subjugated spaces. A glossolaliac invective commentary, inscribed by hand at the drawing's peripheries, supports and focuses the projection of both figures.

New, future corporealities with their axis in projection emerge through acts of anatomical experimentation, in art and film. For Artaud's new, "authentic" body to be projected, its originating and assaulted body must undergo woundings and erasings that will imminently make it disappear, or else cause it to instigate its own vanishing. Artaud's violently conjoined figures – emitting that imminent vanishing, while simultaneously generating a new and insurgent body – are prescient ones for contemporary projects of corporeal projection. Seminal artists of projection – contemporary projectionists – are those who now move beyond the generation of animated moving-image sequences in order to excavate in depth the future configurations of the human anatomy, as it transits engulfing new ocular cultures such as those of digital image-making and is propelled into unprecedented urban environments.

As one of the instigators of Surrealist cinema in the 1920s, and an instigator too in that same era of riots in the auditoria of cinemas, Artaud's intimacy with moving-image projection entailed the compulsion to enact projection's malfunctions and meltdowns. His notebook drawings of the 1940s era of *The Projection of the authentic body* also show isolated projectors, thrown across space or destroyed. Projection's operation demands the incorporation of such malfunctions, as Muybridge always insistently demonstrated. *The Projection of the authentic body* could be seen by its confronted spectator as showing an unconstrained projectionist – maintaining a feral, even murderous isolation – and his conjoined "operator", always bound in servitude as the accomplice of that grandmaster projectionist, as Webster was to Muybridge on their UK tour of rat-infested auditoria, prior to Muybridge's optimal moment of glory during his projections across Europe's cities.

While new corporealities are envisioned in art via an exploratory preoccupation with projection and its spatial dimensions, as in Artaud's (in many ways) cinematic drawing, the preoccupation also with time, especially with time's ricocheting corporeal transits, forms a correspondingly driving obsession of art and film, notably in the temporal configuring of new urban environments via projection's momentums and catastrophes. In Chris Marker's 1962 film *La Jetée*, a prisoner confined within a subterranea below the future war-destroyed city of Paris is induced by his sinister jailers to project himself through time, in a laboratorial experiment with vision, propelled out of the apocalyptic present moment, firstly into the past which strongly adheres him through his memory of a running figure being assassinated on Orly airport's observation pier, and subsequently into the resistant domain of the future, where a new variant of Paris has been created in a spider's web configuration ("ten thousand incomprehensible avenues", as the film's commentary describes it) whose entire survival depends on the success of that prisoner's ongoing experiment in projection. The prisoner's jailers will kill him at the end of his projectionist's itinerant experiment in alternately inhabiting the past and the future, whether that work successfully ensures the future-city's existence or not. The future and past oscillate as entities of ruination and strangeness, and the contaminated present moment is uninhabitable. The prisoner's act of projecting his vision through time forms an exacting, agonising one, without respite and only tenable through his jailers' narcotic injections, as with the sustaining of innumerable other projectionists' work in cinemas' projection-boxes worldwide. Rather than installing himself in the spectacular new Paris once its future is ensured, the prisoner instead projects himself backwards, in a reverse plummeting, into his memory's resilient image of the figure running on the airport's pier; he is immediately killed there by his jailers' assassins as he moves in sequential movement from end to end of the pier, and thereby finally occupies that past figure of his haunted memory.

Marker's moving-image work – which extended across six decades from celluloid film, through video works, to digital media and the inhabitation of "Second Life" digitised cities – also experiments with the sudden transition from the still image to the moving image which formed the heart of Muybridge's projection events, always rendering his spectators awe-struck. *La Jetée* is entirely composed of still images apart from a momentary sequence showing the awakening of the eyes of the woman whom the prisoner encounters in the past, thereby abruptly propelling the film's spectators into the ecstatic zone of moving images and transforming *La Jetée* into a "film" rather than a sequential photographic work. In his preoccupation to engage always with the most current or innovative formats, Marker's work – as with *Lovers* – was often rendered inaccessible by moving-image technology's ongoing rapid mutations, from his video works of the 1980s to his 1990s Hyperstudio-generated CD-ROM *Immemory* with its self-monumentalising of his own past body of work (a project closely paralleling Muybridge's 1894-1904 scrapbook, itself enduringly "readable" in its experimental multi-layered paper-based format), its software now surpassed and unplayable, and to his final 2000s art-focused installation of his work within Second Life cityscapes, prior to that obsolescent medium's shift into a shopping interface.

Muybridge's own mid-projection activation of his work – from the still image to the moving image, that moment delegated to his operator while he himself stood directly alongside his screen – forms an integral part of his own oscillating, non-linear transits in time, with their pivotal malfunctions, extending from that work's capacity to generate astonishment in his European tour's audiences, to its near-oblivion at the moment of his death, and to its multiple inspirations for contemporary artists. In Muybridge's generation of innovative corporealities and immersive urban environments, as in his moving-image projectionist's tours and his exhibitions of his San Francisco panoramas, he also inhabited and transited between locations in a way that annuls spatial coherence, often propelling himself into catastrophic dead-ends as with his Zoopraxographical Hall experiences in Chicago, and fluctuating between phantom or void projections such as those in 1894 Japan and projections so demonstrably packed with spectators, to the point of an acute corporeal overspilling, that the excitement which those projections' unprecedented sensations unleashed is still tangible.

Muybridge's future-oriented corporealities and urban spaces are possessed too by a restless relentlessness and repetition which also drive his projections directly into the contemporary instant. As with urban digitised environments and the perpetually wired figures inhabiting their technological networks and screens, in the present moment, Muybridge's own propulsive projections of his work always formed unremitting ones, in which mishap only ever appeared to incite him further onwards, into new velocities and awry obsessions. The pervasive infiltration into contemporary urban environments of Muybridge's moving-image projections and their screen-located sitings forms a self-evident phenomenon, but Muybridge's image-based bodies too are prescient ones, incessantly in transit, and anticipating bodies now ocularly focused without respite on corporeality's precarious digitised existence, in beguiling image-sequences and raw dances of aberration. That urban corporeality is also one integrally attuned to its own engulfing, terminal malfunction.

The last-surviving projectionist of any city is often an obstinate figure who will not unleash a tight, adroit grip on the projector's Maltese cross, will not suffer expulsion from the projection-box, and whose body will finally need to be carried out screaming, as though it were entirely composed of extremophile organisms, adept at surviving even in the most unliveable and hostile terrain, or in anti-filmic cities. Moving-image projection has held three pre-eminent historical moments of ending, and each forms a vital juncture, retrospectively illuminating projection's instigation and its entire span, within that unique concertinaed instant of closure, which unleashes abandonment and the scattering of cinematic detrita into the spectral future, as well as casting-aside the projectionists. Firstly, Muybridge's Zoopraxographical Hall stands in isolation in his devising and realization of an architectural entity uniquely intended for moving-image projection, in the format of his twelve-inch and

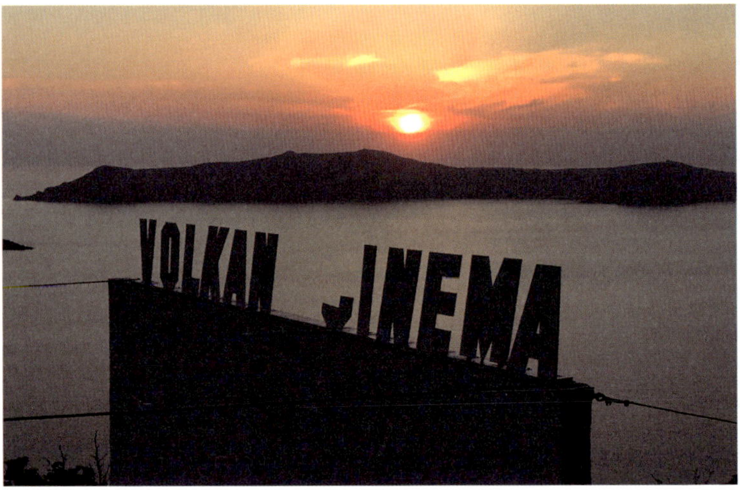

Ill. 24: Volkan Cinema, Santorini, 2020.

sixteen-inch discs; that building's razing configures all future cinemas' razings. Secondly, World's Expositions form multiple locations of experimental moving-image auditoria, pre-eminently through the 1970 Osaka Exposition's domed pavilions intended for immersive-screen projections of artists' corporatised films, experienced by standing or platform-propelled spectators in their millions, enraptured or alienated by that experience; other Expositions, too, presented moving-image experiments such as immense exterior projection screens, and were inhabited by many hundreds of hired projectionists, highly adept but as obsolete as their films and pavilions at the moments of those Expositions' closures, as with that of the Osaka Exposition. Thirdly, the end of celluloid projection entailed the voiding of innumerable cinemas worldwide; celluloid-projection cinemas' pre-eminence extended from 1907-08 to the mid-1990s (the centenary of the Skladanowskys' experimental celluloid-projections, as well as those of their rivals, the Lumière brothers), then entered rapid disintegration across the following decades, rendering obsolete almost all of those cinemas' projectionists in that technological and architectural fall.

The last-ditch celluloid projectionists, habitually now surviving only one to each city, invariably exhibit the enduring obsessions and addictions that sustained their profession's many thousands of now-vanished figures. Projectionists, in their extremity, form still-itinerant bodies, assembling enough work to survive only through combinations of art-cinema projection and pornographic-cinema projection, transiting their cities from one cinema's screenings to another and back. Art-cinemas, often attached to art-museums, resiliently utilise celluloid projection, often in deference to their elderly projectionists' attachments to it, but also to infiltrate cinematic technological histories, via films which palpably resist their digitisation; pornography-cinemas – existing as rarely as the projectionist's own figure, perhaps one to each city and usually located on the urban peripheries – will not expire until their very last gasp, since their patrons require a specifically cinematic experience with its covert spaces and concealing bursts of noise. The figure of the last projectionist is often evoked and lauded in cities' press accounts, as someone who has maintained their profession in isolation for decades, manoeuvring their way through cinemas' ownership shifts as well as via projection's infinite malfunctions. Their exhausted bodies now emanate a city's entangled history. In such press accounts, the projectionist often has no apparent home and may inhabit the projection-box, especially if still engaged in all-night projections; the projectionist may agree to speak only to evoke dead or vanished former colleagues who went "over the edge". Projectionists' vocal reflections on their profession invariably emphasise its dignity, even when their working lives have formed great battles against snapped or burning celluloid and against rushes of dissatisfied spectatorial assailants at their projection-box doors, spitting fury. Above all, those last-breath celluloid projectionists distance themselves from the ascendancy of remote or automated digital projection.

Beyond urban environments, residual locations of open-air celluloid projection survive, in amalgams of the illuminated and the infernal, as at the Volkan Cinema, attached to the sheer caldera cliff-edge on the volcanic island of Santorini. A pathway leads down the cliff-face, and a gate opens out into a narrow terrain of flattened black volcanic soil, with enough space for thirty or so spectators' chairs assembled between the vintage Hokushin 16mm projector on its elongated metal stand, and the screen and its sound-speakers positioned below the cinema's signage, the "C" of its "Cinema" subsisting only as a fragment. Night must fall, and the adjacent volcanic islands fade out, before the outdoor projection can begin on that cinematic precipice, so the spectators sit as though in petrifaction, until they feel the mysterious figure of the projectionist slip silently into position behind them and operate the projector to beguile them, with the sudden apparition of film on that last-ditch screen. That cinema's sense of spectatorial exposure resonates with the past experiences of lunatic asylums' outdoor-cinemas, prevalent in France and Italy in the mid-twentieth century decades, installed in their institutions' surrounding parkland, where the patients also sat waiting in frozen excitement for the projectionist to initiate the projection, often of Dreyer's *The Passion of Joan of Arc*, seen in repetition for the second or hundredth time; further back in outdoor celluloid-projection's history, itinerant projectionists' conjuration of film in village squares (Jean Genet's first contact with film, in his 1910s village in France's Morvan region) also immersed spectators in their sensory experience's precipice, caught between illuminated ecstasy and a plummeting fall.

By contrast with those illuminated contacts with the last-surviving outdoor celluloid projectionists' work, as at the Volkan Cinema, the inhabitation of digitised cinemas may appear to transmit an oblivious darkening of cinematic space. That sense of an indefinite plunging into darkness is especially acute at cinemas whose automated digital projection breaks down or whose near-obsolete projectionist forgets to turn the cinema's lights back on at the film's end (that multi-tasking projectionist may be sitting in the cinema's bar, operating the projection via a misfiring iPhone, or else from the ticket-booth, its status now conjoined with that of the redundant projection-box). At an afternoon screening at Tangier's Rif cinema, of Godard's *Le Mépris*, the audience gradually realised, after the film's final shots of glaring Mediterranean marine-light, in the subsequent post-film darkness, that the auditorium's lights would never be switched on, and so had to grope in transit through the darkness, body against body, to find the cinema's emergency exit and be propelled out once again into glaring marine-light, of the kind that Muybridge may have imagined that he could conjure up – refracted obliquely from Lake Michigan, from the time when he had left the Chicago Exposition's grounds at the summer's end in 1893, or else from California, whose coastline he had mapped in an 1871 photographic project – for his moment of death in 1904 while engaged in arduously digging his back-garden's swimming-holes.

Filmmakers still stand in corporeal exposure beside their projection screens – as Muybridge stood alongside his own screen (whose large format, accentuating his moving-image sequences' clarity, he repeatedly extolled), as at the Berlin Urania, Munich Odeon and Zoopraxographical Hall – as though replicating that formative proximity of the projectionist's body and the corporeally-inhabited screen. That vocal act in intimacy with the screen can take the form of an irresistible revisiting of projection's imperatives, resonating with the driving repetition which Muybridge himself instilled in his vocal commentaries, or else it may articulate an elegiac final-breath valediction directed at the last-surviving, expiring projectionists. At a screening at Berlin's Wolf-Kino in 2020 of his 2002 work *Decasia*, on film's disintegration and abandonment, the filmmaker Bill Morrison positioned himself alongside the screen in order to maintain a vocal commentary on that film of incendiary-prone archival celluloid fragments, transferred by Morrison to a digital format that appeared only momentarily to be able to contain their conflagrations, in film's end. Morrison concluded his commentary: "Film can't survive, digital media can't survive." – And after a long silence, he continued: "Only the act of projection survives."

Muybridge was an inhabitant of cities, determined to render their dimensions in panoramic and other sequential photographic forms, as well as an itinerant transiter of cities, as on his Central European moving-image projection tour. In his inhabitations of San Francisco, Philadelphia and the Chicago Exposition's experimental city, Muybridge remained perpetually exposed to the expansive urban technological experimentation of the USA's 1870s-90s era, such as its rapid developments in transport networks from city-connecting continental railways to inner-city systems, data and media innovations such as those of photography's incorporation into print media and telegraphy's movement into wireless telegraphy, and architectural innovations in multi-storey and subterranean construction, among many other areas. Muybridge experienced cities in an extreme flux that parallels and anticipates contemporary cities' relentless transformations.

Muybridge's distinctive urban focus was on San Francisco, from his work in street-level photography through his "Helios Flying Studio" (whose promotional icon was a winged camera, accompanied by a written commentary guaranteeing "perfect satisfaction" to whoever commissioned Muybridge to photograph and then immediately develop his images of any urban location) to his experimentation in panoramas photographed from San Francisco's elevations; by contrast, he appears never to have been an opportunistic documenter of cities, and did not photograph the cities of Central Europe which he traversed in 1891, nor the Chicago Exposition site, nor Kingston-on-Thames. Philadelphia's urban presence is glimpsed only in fragments at the far edges of Muybridge's sequential photography of bodies and animals undertaken in university yards and civic zoos. Always acutely receptive to malfunction and its catastrophes, Muybridge may presciently have photographed San Francisco with such intensity in anticipation of its large-scale seismic incineration and vanishing in 1906, two years after his death. As with the traumatised ex-detective Scottie Ferguson's incessant corporeal transits of San Francisco in Hitchcock's film *Vertigo*, as Ferguson searches for traces of the identity of the mysterious figure of Madeleine and oscillates precariously between the future, present and past, Muybridge's own photographic trajectories through San Francisco with his "flying" camera appear as the anticipatory mapping of its cityscape for locations of his wife's covert encounters with her lover, still five years or so in the future, and the pinning-down of the location from which he could rapidly depart that city for his short journey northwards, by ferry and train, to Calistoga, to kill that lover with a single shot to the heart, Muybridge's adroit hands already finely attuned to rapid, split-second gestures, as with those of all future projectionists.

When Muybridge exhibited his 1877 San Francisco eleven-panel horizontally-formatted urban panorama (the first of two he undertook in 1877-78, with the second panorama's panels executed in a vertical format) as an immersive environment at the city's Art Association galleries, its spectators, once their initial disorientation and vertigo subsided, would have oscillated in their ocular perception of its sequential panels, between examining minuscule urban details and scanning its immense span, encompassing dense central streets' areas of near-darkness and the illuminations of its perimeters' marine light. Time is veering in the panorama's elongated spatial simultaneity; clockfaces visible in the panorama (which took around five hours to accomplish) show times wrenched far apart. In leaving that image-generated immersive environment, and stepping out of the Art Association building, those spectators would then have re-entered San Francisco's immediate live-action environment of bodies and cacophonies, that city's time now propelled infinitesimally onwards into the future from its panoramic rendering. Muybridge was enduringly attached to the urban experimentation of his panoramas of San Francisco. Several press accounts of his 1891 tour of Europe's cities, such as the

Wiener Zeitung's account of his Vienna Technical High School event, suggest that he still occasionally showed elements of one or other of his San Francisco panoramas, during his projections' last phase which appears to have taken a semi-improvisational approach to its content, often including Muybridge's projections of naked figures, once he had been able to ascertain whether or not those sequences would scandalise a particular audience; the presence of the San Francisco panoramas very probably appeared in the form of still images shown from a slide-projector, rather than being projected from Muybridge's Zoopraxiscope. The acute sensory oscillations in spatial dimensions and in time generated for its spectators through the immersive installation of Muybridge's 1877 San Francisco panorama experiment are still tangible whenever one or other of the original panoramas (such as Muybridge's own set of prints of the 1878 panorama, which he maintained with him until his death) is installed in space, to surround its spectators, as in the galleries of the Tate Britain exhibition devoted to Muybridge's work in 2010-11.

Inhabitants of the plazas of contemporary digitised cities, gripping their smartphones and immersively surrounded by corporate towers holding immense moving-image screens, undergo urban perceptual experiences initiated and determined by Muybridge's experiments in the panorama medium and in the screen-projection for spectators of moving-image sequences focused on rapid corporeal movement. The occupants of contemporary cities are simultaneously spectators and projectionists. They must incessantly adjust their ocular scale, from a focus on their smartphones' minuscule screens, to the maximal expanse of the surrounding city with its preeminent constellations of digital moving-image screens, in eye-recalibrations alongside corporeal gestures of contortion and rotation that were also integral to spectators' experiences of the Osaka Exposition pavilions' vast-scale immersive projections. That perpetually self-choreographed ocular and gestural oscillation between smartphone screens and urban towers' image-screens itself forms the work of a spectacular projectionist, synchronising spatial image-sources in order to generate a maximal sensory self-engulfing; urban projectionists project to themselves, in what appears at first sight to be an autonomous act, but is immediately striated and disintegrated by its controlling corporate powers' imperatives and overseeings.

The intensive work of urban inhabitants' incessant gestural convulsion and rapid ocular oscillation, between minuscule and maximal screens, as the contemporary experience of urban space, will eventually generate new corporealities and new eyes, each possessing variants of an aberrant outlandishness that Muybridge seized in his motion-photography sequences and then projected onto his own moving-image screens, and which Francis Bacon subsequently accentuated in his paintings of malfunctioned, collapsed bodies. New anatomical capacities and reconfigurations are demanded for those extreme sensory and corporeal manoeuvres undertaken between urban space and screen space, and future cities may well be increasingly populated by figures clearly identifiable from the distinctive permutations of Muybridge's experiments. Muybridge himself embodied an obsessionally driven figure in his work, preoccupied relentlessly with repetition's sensorial intensities: a precursor and animator of the contemporary city's corporeality. The experience of multiple ruptures, splits and severings of perception, in contemporary urban space, is already one profoundly familiar and integral to moving-image projection and its histories, and to the work of the projectionists.

At Muybridge's Zoopraxographical Hall on the Midway Plaisance, the textually-inscribed hoardings positioned outside, in the form of two large boards leaning against the central pillars of the auditorium's entrance and facing out to the public gaze, were designed to attract the ocular attention of passers-by and propel them inside; Hartmann's photograph vividly catches an instant of that attraction in operation, as the two women in white and black dresses are arrested in their movement by the hoardings' beguiling textual commentary on Muy-

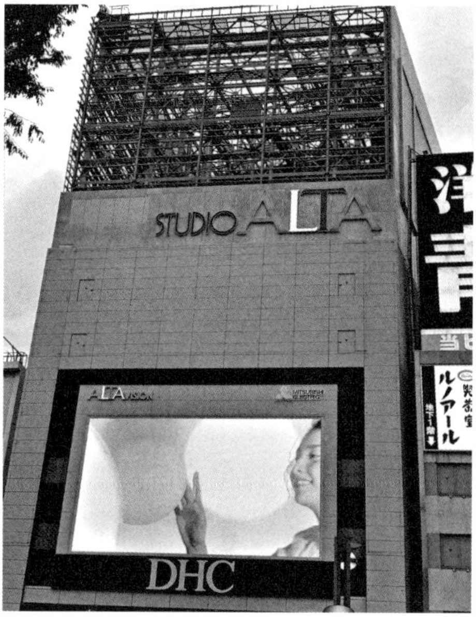

Ill. 25: Studio Alta urban screens, active and malfunctioned, Tokyo, 2020.

bridge's projections, and are clearly now deliberating whether to go inside, transiting the Zoopraxographical Hall's entrance (as two of the approximately 1,280 spectators to do so) to experience those projections and Muybridge's live-action commentary on them.

Prior to their contemporary animation by moving-image sequences, the immense digital screens of urban plazas – viewed by their inhabitants in acute oscillation with the miniaturised screens of handheld devices – were similarly occupied by beguiling textual content, in the era of neon screens. The facade and interior of Muybridge's Zoopraxographical Hall form an accelerated preview of the history of corporate urban screens attached to department-store or hotel towers in worldwide metropolises. Innovations in the technologies of neon hoardings in the early 1960s allowed them to constitute vertically installed screens, positioned almost directly alongside one another to form vastly immersive illuminated environments surrounding plazas such as Shinjuku Station's East-Exit in Tokyo and the Alexanderplatz in East Berlin. The neon hoardings often also displayed icons of products, but their content was predominantly textual, in exclamatory commentaries that could consist solely of the repeated name of a corporation, projected succinctly to be memorised (Toshiba, Fuji) by urban spectators, as in Shinjuku, or else extolled state industries, such as the German Democratic Republic's plastics industry, in the Alexanderplatz. Those neon screens' spectators were induced to enter, either literally into the department-store entrances below the screens, or else into a condition of subjugated complicity with corporate or state-industry imperatives. The screens, in their spectacular projections of neon-animated letterings' colours, were documented in urban filmmakers' work, as in the films of Nagisa Oshima shot in Shinjuku, and in films by cinematographers located in the Alexanderplatz, often tracking the movements into that plaza of youth-festival processions emerging from the adjacent Karl-Marx-Allee. But by the 1980s and early 1990s, with the innovation in Japan of large-format video image screens such as those presented at the Tsukuba Exposition (almost-immediately obsolete and transformed into projection's detrita, to appear as Miyamoto's photographed "architectural apocalypses"), and the dissolution of the German Democratic Republic state-industries' focus in the vanished East Berlin, those commentary-inscribed neon screens were consigned to oblivion, their few survivors confined in museums or archives as artefacts of defunct technologies.

The contemporary digital screens which replaced them by the 2000s, perpetually upgraded to the 2020s and magnified in size and resolution, incorporate no commentary, as though their urban spectators can effortlessly be propelled without any transitional instant of enticement into their zones of corporeal movement, as the audiences for Muybridge's Zoopraxographical Hall would have been if they had displayed the extreme zeal Muybridge had anticipated, with his adding to his architectural design of extra kiosks to facilitate the entrance of teeming, overspilling crowds such as those he had attracted in Berlin, Munich and Vienna. Instead, Muybridge's Chicago-Exposition potential spectators displayed hesitancy or reluctance, in the way that contemporary urban spectators, disabused with corporate imperatives, and preoccupied instead with ecological, activist or nihilistic imperatives, often display reluctance to their enmeshing within urban digital enthralments, thereby intimating that those environments' screens will suffer a future fall, into malfunction and obsolescence. Those screens' pervasive digital animations of abbreviated, urgent and corporeally-focused sequences – designed to be assimilated during spectators' brief transits through plazas, on foot or in vehicles, and operating through incessant repetition – intersect intimately with those projected in his auditoria by Muybridge, of naked or near-naked figures walking and abruptly turning in acts of outlandish corporeal exposure, thereby conjuring the salutary delusion, in exhausted dead-of-night urban traversals, that Muybridge himself were the enduring projectionist, as well as the instigator, of all contemporary urban digital cityscapes, without exception.

On one of those dead-of-night urban walks, through Shinjuku's avenues, after a day of investigating Muybridge's 1894 Tokyo phantom projections in that city's archives, I entered the East-Exit plaza's immersive environment of huge screens projecting corporeal sequences from department-store towers, and saw that one, positioned on the facade on the first-ever Shinjuku department-store to have installed a digital screen, the Studio Alta, had catastrophically malfunctioned, while the screen positioned below it still projected its sequences. Fragments of the self-exploded screen had scattered on the pavement. On the next day, in daylight, spectators watched, many weeping, as technicians on hydraulic gantries dismantled the shards of the blown-out screen. I asked the ground-level technicians what had happened, but bewildered, they knew nothing: it was a mystery of projection, as though that screen had negated itself in mid-projection, in an all-engulfing breakdown. Their occupation now was that of anti-projectionists, undoing projection's work. On the following day, that dismantling had been completed and the abandoned, denuded screen remained there, at the Studio Alta facade's summit, its twenty panels' metal carapace rawly exposed, the screen's terminal malfunction accentuated by its proximity to the still-functioning lower screen projecting a sequence showing a woman's illuminated face as she transited that screen's space. The blown-out screen was visible but voided, as though to demonstrate – to all spectators in the plaza below – moving-image culture's wholesale annulling in its exhausted digitised corporate contamination, in a necessary stand-still moment before the moving image's future re-activation in an unprecedented form, as an anatomical and ocular reconfiguration, at the adroit hands of an omniscient instigator, as though as the very last of Muybridge's phantom projections. By my final day in Tokyo, the Shinjuku crowds no longer looked upwards at that malfunctioned screen and appeared even to have forgotten about its existence.

Muybridge's projections vitally anticipate all future moving-image projections in all media of the contemporary moment, and his own initiating work as a projectionist, in transit through Europe's cities and installed in his own projection auditorium at the Chicago World's Columbian Exposition, prefigures all future projectionists' work and their often-obstinate, aberrant preoccupations. The survival of Muybridge's work appears simultaneously pervasive and fragile. After a projection's short-lived inhabitation of a moving-image screen, its delicate detrita endure a terminal deliquescence as they absorb contaminants from the surrounding atmosphere, as malfunctioned fragments closely paralleling those of a corporeal disassembling. The anatomical dissolution of projection itself presents a spectacular ocular and sensory experience. Remnants of projection, as with the traces of Muybridge's projections, often overspill the media that attempt to contain them, as Muybridge's European-tour audiences overspilled his auditoria. Eventually, projection and the projectionists will enact a fall, into their vanishing, but remain always resiliently ready to erupt and insurge again.

Bibliography

This book was primarily researched in a wide range of archives and collections, and through explorations of many urban sites, in Europe, the USA and Japan; this bibliography aims solely to assemble the relatively small number of publications relevant to the book.

Applebaum, Norman, *Spectacle in the White City*, Calla Publishers, Mineola, 2009.

Braun, Marta, *Eadweard Muybridge*, Reaktion, London, 2010.

Brown, Julie K., *Contesting Images: Photography and the World's Columbian Exposition*, University of Arizona Press, Tucson, 1994.

Castan, Joachim, *Max Skladanowsky, oder der Beginn einer deutschen Filmgeschichte*, Füsslin Verlag, Stuttgart, 1995.

Derrider, Jean-Paul, *Cinema*, Hatje Cantz Verlag, Ostfildern, 2007. Derrider's 2002-07 photographic essay evokes the final cinema projectionists of Germany and Belgium: "Düsseldorf, January 2002, the indestructible B12 [projector] is giving off sparks. The projectionist appears suddenly. He cuts the power and reappears with his soldering iron, opens the lamphouse door on the Bauer projector and, after a quick and efficient repair, restarts the projection." (page 7) Derrider records projectionists' laconic vocal commentaries upon their profession: "I wanted to be a projectionist. Once I went in, I never came back out."... "Cinema is just addictive." (pages 12, 20)

Eckersall, Peter, *Performativity and Event in 1960s Japan: City, Body, Memory*, Palgrave Macmillan, London, 2013.

Expo '70: Fuji Group Pavilion, Exposition programme, Fuji Group Corporation, Tokyo, 1970.

Expo '70: Midori Pavilion, Exposition programme, Midori-kai Corporation, Ōsaka, 1970.

Expo '70: Pepsi Pavilion, Exposition programme, Pepsi Corporation, New York/Tokyo, 1970.

Expo '70: Toshiba IHI Pavilion, Exposition programme, Toshiba Corporation, Tokyo, 1970.

Haas, Robert Bartlett, *Muybridge: Man in Motion*, University of California Press, Berkeley, 1976.

Holborn, Mark, *Beyond Japan: A Photo Theatre*, Jonathan Cape, London, 1991.

Japan World Exposition, Osaka, 1970: Official Report, Commemorative Association for the Japan World Exposition (1970), Osaka, 1971.

Marker, Chris, *La Jetée*, Zone Books, New York, 1996.

Marker, Chris, *Immemory*, Exact Change, Boston, 1997/2008.

Mroué, Rabih, *Image(s), mon amour: FABRICATIONS*, Centre de Arte Dos de Mayo, Madrid, 2013.

Muybridge, Eadweard, *Descriptive Zoopraxography*, Exposition programme, University of Pennsylvania, Philadelphia, 1893.

Ralph, Julian, *Chicago and the World's Fair*, Harper and Brothers, New York, 1893.

Sinkevitch, Alice (ed.), *AIA Guide to Chicago* (second edition), Harcourt Inc., Orlando, 2004.

Solnit, Rebecca, *River of Shadows: Eadweard Muybridge and the Technological Wild West*, Viking, New York, 2004.

Sylvester, David, *The Brutality of Fact: Interviews with Francis Bacon*, Thames and Hudson, London, 1987.

Thévenin, Paule and Derrida, Jacques, *Antonin Artaud: dessins et portraits*, Gallimard, Paris, 1986.

Völschow, Undine, *Nachlass Max Skladanowsky*, Bundesarchiv, Koblenz, 1995.

Illustrations List

1. *42nd Street movie theater audience, N.Y.C.*, 1958 (Diane Arbus; British Film Institute).
2. *Man working with a projector*, 1958 (Marion S. Trikosko; Library of Congress).
3. Eadweard Muybridge, letter fragment, 15 July 1891 (University Archives and Records Center, University of Pennsylvania).
4. Muybridge's Berlin Urania auditorium, now an abandoned auditorium of the Moabit police compound, 1, 2020 (Holger Hartung; Free University Berlin).
5. Muybridge's Berlin Urania auditorium, 2, 2020 (Stephen Barber).
6. Muybridge's Munich Odeon auditorium, now an interior courtyard of the Bavarian Ministry of the Interior, Sport and Integration, 2020 (Stephen Barber).
7. Muybridge's Vienna auditorium, at the *Technische Universität*, 2020 (Stephen Barber).
8. Muybridge's Vienna auditorium, at the *Künstlerhaus*, 2020 (Stephen Barber).
9. Untitled photograph of Muybridge's "Zoopraxographical Hall" under construction at the Chicago World's Columbian Exposition site, 1892-93 (unknown photographer; World's Columbian Exposition Collection, Chicago History Museum).
10. Untitled photograph of figures walking on the Midway Plaisance avenue at the Chicago World's Columbian Exposition, in front of Muybridge's "Zoopraxographical Hall", 1893 (Joseph E. Hartmann; World's Columbian Exposition Collection, Chicago History Museum).
11. The Lange House of Muybridge's architect Thomas W. Wing (1893), 2020 (Stephen Barber).
12. The Raber House of Muybridge's architect Thomas W. Wing (1870/ renovated c.1892, now derelict), 2020 (Stephen Barber/Jeremy Biles).